Contents

To the student   v
Acknowledgements   vi

1 HUMANS AS ORGANISMS

1.1	Working together	2	1.16	Fighting disease	32
1.2	Muscles in action	4	1.17	Lose it	34
1.3	A living framework	6	1.18	A filter for life	36
1.4	A winning smile	8	1.19	Feeling sweaty	38
1.5	Food is important	10	1.20	Staying warm	40
1.6	It's your choice	12	1.21	Keeping a balance	42
1.7	Processing food	14	1.22	Life in the balance	44
1.8	Enzymes at work	16	1.23	Fast track	46
1.9	Breath of life	18	1.24	Use your senses	48
1.10	Breathing all your life	20	1.25	A1 vision	50
1.11	Hale and hearty	22	1.26	Passing the message	52
1.12	Living pump	24	1.27	Drugs can harm	54
1.13	Blood networks	26	1.28	Everyday harm?	56
1.14	Life blood	28	Exam questions		58
1.15	Invaders	30			

2 PLANT LIFE

2.1	Growing for gold	62	2.9	Water for life	84
2.2	A good start	64	2.10	Supporting plants	86
2.3	Measuring growth	68	2.11	Water uptake	88
2.4	Plants are producers	70	2.12	Losing water	90
2.5	Food factories	74	2.13	On the move	92
2.6	More and more	76	2.14	Controlled growth	94
2.7	Feeding plants	78	2.15	Changing nature's way	96
2.8	Respiration matters	82	Exam questions		98

3 THE ENVIRONMENT

3.1	It's our environment	102	3.12	Decay for renewal	124
3.2	Places to live	104	3.13	Using decay	126
3.3	Matching the conditions	106	3.14	Round and round	128
3.4	Using keys	108	3.15	Nitrogen for protein	130
3.5	Studying habitats	110	3.16	Pesticides	132
3.6	Living together	112	3.17	The organic way	134
3.7	Food webs	114	3.18	Pollution	136
3.8	Energy transfer	116	3.19	Pollution control	138
3.9	Community relations	118	3.20	In the air	140
3.10	Dynamic numbers	120	3.21	Working with ecosystems	142
3.11	Managing the environment	122	3.22	Our world	146
			Exam questions		148

4 REPRODUCTION, INHERITANCE AND EVOLUTION

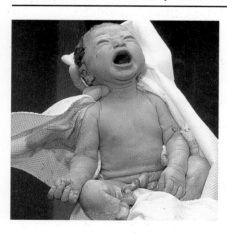

4.1	Sorting organisms out	152	4.10	Breeding to order	174
4.2	It's all changed	154	4.11	Human reproduction	176
4.3	Selection and evolution	158	4.12	Planned pregnancy	178
4.4	We're all different	160	4.13	Controlling fertility	180
4.5	Reproduction	162	4.14	Passing it on	182
4.6	Two kinds of division	164	4.15	Inheriting features	184
4.7	Plant enterprise	168	4.16	What's the chance?	186
4.8	All the same	170	4.17	Changing the message	188
4.9	More than a flower	172	4.18	Mutation	190
			Exam questions		194

Glossary 196
Index 200

BIOLOGY

S Gater   V Wood-Robinson   Series editor K Foulds

Other titles in this series:
GCSE Science Double Award Chemistry ISBN 0 7195 7158 8
GCSE Science Double Award Physics ISBN 0 7195 7159 6

© S Gater, V Wood-Robinson, K Foulds, 1996

First published in 1996 by
John Murray (Publishers) Ltd
50 Albemarle Street
London W1X 4BD

Reprinted 1997

All rights reserved. No part of this publication may be reproduced in any material form (including photocopying or storing in any medium by electronic means and whether or not transiently or incidentally to some other use of this publication) without the written permission of the publisher, except in accordance with the provisions of the Copyright, Designs and Patents Act 1988 or under the terms of a licence issued by the Copyright Licensing Agency.

Artwork by Art Construction, Peter Bull Art Studio, David Farris and Technical Art Services.
Layouts by Can Do Design.
Typeset in Rockwell Light and News Gothic by Wearset, Boldon, Tyne and Wear.
Printed and bound in Great Britain by Butler & Tanner Limited, Frome and London.
A CIP catalogue record for this book is available from the British Library.

ISBN 0 7195 7157 X

To the student

This textbook has been written to help you study biology as part of a science course towards GCSE assessment. It includes details to cover the GCSE syllabus of many examination boards and may provide more detail than you will need for your examination. So, you should check carefully with your teacher whether there are any parts that are less relevant to your studies.

The book is divided into four sections, each covering a major theme of biology. Within each section most topics are written at foundation level. You should read these topics fully. The remaining topics are written at a higher level; you may not need to study all of these.

Each topic provides background information and explanation set in different contexts, allowing you to think through the main ideas fully. Exercises draw upon information provided in a topic, so you may answer questions to show an understanding, follow instructions to carry out practical work and investigate your own ideas.

Any book can give only a certain amount of information, so you should try to develop confidence in searching for information from a variety of sources both in and out of school. This textbook provides a stimulus to find out more about biology – use your curiosity to seek out further data, to increase your understanding and to inform your investigations.

Examination questions are given at the end of each section, so you can test your knowledge and understanding on a regular basis. Higher level questions are denoted by a solid bar. The glossary could also be used as a self-test or as a simple reference. Try to read and re-read the relevant topics often so you become familiar with the ideas and learn effectively. Frequent revision will improve your understanding and recall of facts – and this will make further learning easier!

Finally, enjoy your studies. Biology is a fascinating collection of facts and concepts, underpinned by systematic scientific investigation. It is one aspect of science, one way to look at ourselves and the world in which we live. So find out more about yourself and the other forms of life, past and present, that make our world such a rich place to live in!

Acknowledgements

The authors wish to thank their families, friends, students at Washington School and staff at John Murray for their huge support, patience and encouragement during the gestation of this book.

Acknowledgements

The authors and publishers are grateful to Mr Vernon Hudson, Senior Teacher and Head of Science at Radcliffe High School, for his advice on the texts of GCSE Science Double Award Biology, Chemistry and Physics.

Exam questions are reproduced by kind permission of:
the Midland Examining Group (MEG);
Northern Examinations and Assessment Board (NEAB);
Southern Examining Group (SEG);
University of London Examinations and Assessment Council (ULEAC);
Welsh Joint Education Committee (WJEC).

The Publishers have made every effort to trace copyright holders, but if they have inadvertently overlooked any they will be pleased to make the necessary arrangements at the earliest opportunity.

Photo credits

Cover: (centre) Waterfall, Blue Mountains, New South Wales, Australia; (left) Antarctic ice formations; (right) Sand ripples in Monument Valley, Utah, USA. All courtesy of ZEFA. **p.iii** t ZEFA, b Inge Spence/Holt Studios; **p.iv** t Laurie Campbell/NHPA, b Keith/Custom Medical Stock Photo/Science Photo Library; **p.1** ZEFA; **p.2** tl Lupe Cunha, tr Thames Water Utilities Limited, cl John Downer/Planet Earth Pictures, cr ZEFA-IDEM, M., bl Heather Angel, br ZEFA; **p.4** ZEFA; **p.6** l Shout Pictures, r Science Photo Library; **p.7** Andrew Syred/Science Photo Library; **p.8** Andrew Lambert; **p.9** l Science Photo Library, r Don Mackean; **p.12** Tony Craddock/Science Photo Library **p.14** l Ecoscene/Martin Jones, c NASA/Science Photo Library, r Shout Pictures; **p.16** John Townson/Creation; **p.17** l, c, r Last Resort; **p.18** Action Images; **p.20** John Cleare/Mountain Camera; **p.22** tl Mike Powell/Allsport UK, tr ZEFA, b Simon Fraser/Science Photo Library; **p.25** Science Source/Science Photo Library; **p.27** ZEFA; **p.28** l Jerry Mason/Science Photo Library, r Shout Pictures; **p.29** Bill Loncore/Science Photo Library; **p.30** l and r M.I. Walker/Science Photo Library, c M. Wurtz/Biozentrum, University of Basel/Science Photo Library; **p.31** Secchi-Lecaque/Roussel-UCLAF/CNRI/Science Photo Library; **p.33** National Library of Medicine/Science Photo Library; **p.36** t Astrid & Hahns-Frieder Michler/Science Photo Library, b Hank Morgan/Science Photo Library; **p.39** Natrel Plus is a registered trademark of the Gillette Company; **p.41** tl Stephen Dalton/NHPA, tr Andy Purcell/ICCE, b Heather Angel; **p.42** Mark Clarke/Science Photo Library; **p.46** ZEFA; **p.48** Shout Pictures; **p.51** t and c Andrew Lambert, b Nikki Taylor; **p.52** Last Resort; **p.53** BSIP VEM/Science Photo Library; **p.54** t Clive Dixon/Rex Features, b Home Office; **p.55** Department of Health; **p.56** Department of Transport; **p.61** Inge Spence/Holt Studios; **p.62** l John Townson/Creation, r The Harry Smith Horticultural Photographic Collection; **p.64** tl, tcr, bl, bcr Sutton Seeds, tcl, tr, bcl, br John Townson/Creation; **p.68** tl Martin King/Planet Earth Pictures, tc Heather Angel, tr The Harry Smith Horticultural Photographic Collection, b Heather Angel; **p.69** London Scientific Films/Oxford Scientific Films; **p.70** John Durham/Science Photo Library; **p.71** Bob Gibbons/Holt Studios; **p.72** Heather Angel; **p.73** Hans Christian Heap/Seaphot Limited/Planet Earth Pictures; **p.74** l Biophoto Associates, r Steve Gater; **p.75** John Townson/Creation; **p.76** Inge Spence/Holt Studios; **p.78** t Miracle Garden Care Ltd, bl, bcl, bcr, br Nigel Cattlin/Holt Studios; **p.79** l Daniel Heuclin/NHPA, r Duncan McEwan/BBC Natural History Unit; **p.81** John Townson/Creation; **p.82** t Nigel Cattlin/Holt Studios, b Ecoscene/Anon; **p.83** CEPHAS/Stuart Boreham; **p.84** t Miracle Garden Care Ltd, b Nigel Cattlin/Holt Studios; **p.85** t Nigel Cattlin/Holt Studios, bl The Harry Smith Horticultural Photographic Collection, bcl Heather Angel, br John Lee/Planet Earth Pictures, br Pedigree Petfoods; **p.86** l and c J.C. Revy/Science Photo Library, r Rosemary Mayer/Holt Studios; **p.87** Alfred Pasieka/Science Photo Library; **p.90** Manfred Kage/Science Photo Library; **p.91** Ernie James/NHPA; **p.94** Heather Angel; **p.96** Elizabeth MacAndrew/NHPA; **p.97** t and b Nigel Cattlin/Holt Studios; **p.101** Laurie Campbell/NHPA; **p.102** tl Andrew D.R. Brown/Ecoscene, tr Ian Took/Biofotos, bl G.I. Bernard/NHPA, br B & C Alexander/NHPA; **p.103** fig. a Gordon Roberts/Holt Studios, fig. b Martin Jones/Ecoscene, fig. c Richard Janulewicz/Ecoscene, fig. d David Woodfall/NHPA, fig. e Sally Morgan/Ecoscene, fig. f Photograph courtesy of British Petroleum; **p.104** (clockwise from top left) Jane Gifford/NHPA, Heather Angel, David Woodfall/NHPA, Heather Angel, David Woodfall/NHPA, John Shaw/NHPA, David Woodfall/NHPA; **p.106** tl and tr Laurie Campbell/NHPA, cl Stephen Dalton/NHPA, cr G.I. Bernard/NHPA, b David Woodfall/NHPA; **p.111** David Woodfall/NHPA; **p.112** l and r ZEFA; **p.114** Heather Angel; **p.115** Christophe Ratier/NHPA; **p.117** l Nigel Cattlin/Holt Studios, r Inge Spence/Holt Studios; **p.118** t G.I. Bernard/NPHA, cl Laurie Campbell/NHPA, cr Melvin Grey/NHPA, b Simon King/BBC Natural History Unit; **p.119** Stephen Dalton/NHPA; **p.120** Heather Angel; **p.122** t Schaffer/Ecoscene, bl Forest Life Picture Library, br Daniel Heuclin/NHPA; **p.123** l Julie Meech/NHPA, r Manfred Danegger/NHPA; **p.124** Heather Angel; **p.125** Nigel Cattlin/Holt Studios; **p.126** Blackwall; **p.128** t Sally Morgan/Ecoscene, b Heather Angel; **p.129** Heather Angel; **p.131** t Sally Morgan/Ecoscene, b Dr Jeremy Burgess/Science Photo Library; **p.133** t Brian Hawkes/NHPA, b Jason Venus/Biofotos; **p.134** l Bob Gibbons/Holt Studios, r R.T. French/Planet Earth Pictures; **p.135** tl Rosemary Meyer/Holt Studios, tr Gordon Roberts/Holt Studios, b Heather Angel; **p.136** l Janet & Colin Bord/Fortean Picture Library, r Nigel Cattlin/Holt Studios; **p.137** t Chinch Gryniewicz/Ecoscene, b Ian Harwood/Ecoscene; **p.139** t Francis Bunker, b Vanessa Vick/Science Photo Library; **p.140** NOAA/Science Photo Library; **p.142** tl Heather Angel, tr Chrissie Houghton/Oxford Scientific Films, b Cooper/Ecoscene; **p.143** Popperfoto; **p.144** tl John Mead/Science Photo Library, r Martin Bond/Science Photo Library, bl David Woodfall/NHPA; **p.145** t and b Mark Boulton/ICCE; **p.146** David Walker/Ecoscene; **p.151** Keith/Custom Medical Stock Photo/Science Photo Library; **p.152** t J. Sainsbury plc, fig. a G.J. Cambridge/NHPA, figs b and c Heather Angel, fig. d The Sun/Rex Features; **p.153** figs a, b and c Heather Angel; **p.154** t George Bernard/Science Photo Library, c Philippe Plailly/Eurelios/Science Photo Library, b Ancient Art & Architecture Collection; **p.155** tl The Natural History Museum, London, tr Tony Craddock/Science Photo Library, bl C. Gilbert/British Antarctic Survey, br The British Museum; **p.156** (clockwise from top left) Sinclair Stammers/Science Photo Library, The Natural History Museum, London, M. Wurtz/Biozentrum, University of Basel/Science Photo Library, The Natural History Museum, London, The Natural History Museum, London, John Reader/Science Photo Library, George Barnard/Science Photo Library, The Natural History Museum, London, The Natural History Museum, London; **p.157** Alfred Pasieka/Science Photo Library; **p.158** The Natural History Museum, London; **p.159** t Michael Tweedie/NHPA, b Heather Angel; **p.160** tl Heather Angel, tr The Harry Smith Horticultural Photographic Collection, bl ZEFA, br Chris Priest/Science Photo Library; **p.162** l Dr Kari Lounatmaa/Science Photo Library, r ZEFA; **p.164** Hattie Young/Science Photo Library; **p.167** Zeneca Seeds; **p.169** G.I. Bernard/NHPA; **p.171** Biophoto Associates; **p.172** tl John Shaw/NHPA, tc Heather Angel, tr David Currey/NHPA, b The Harry Smith Horticultural Photographic Collection; **p.173** t G.J. Cambridge/NHPA, b Heather Angel; **p.174** Sutton Seeds; **p.175** t E.A. James/NHPA, b Brian Hawkes/NHPA; **p.176** t Keith/Custom Medical Stock Photo/Science Photo Library, b David Scharf/Science Photo Library; **p.177** BSIP VEM/Science Photo Library; **p.182** l ZEFA, r Andrew Lambert; **p.184** Andrew Lambert; **p.186** Simon Fraser/RVI, Newcastle-upon-Tyne/Science Photo Library; **p.187** l Andrew Syred/Science Photo Library, r Science Source/Science Photo Library; **p.188** James Holmes/Celltech Ltd/Science Photo Library; **p.189** Sinclair Stammers/Science Photo Library; **p.190** Hulton Deutsch Collection; **p.191** Philippe Plailly/Science Photo Library; **p.193** Rob Stratton.

(t = top, b = bottom, r = right, l = left, c = centre)

1
HUMANS AS ORGANISMS

1.1 Working together

Humans, like other animals and plants, must do certain things to keep themselves alive and to enable the human race to survive. Some features of what we do are common to all living things.

Living things survive by **reproduction**

Living things must get rid of waste products. **Excretion** gets rid of waste products made by the body. Sewage treatment processes human wastes

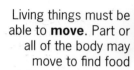
Living things must be able to **move**. Part or all of the body may move to find food

Living things need **food**. It supplies chemicals such as calcium (needed for bone, and others that supply energy. We need to eat food. Plants make their own by **photosynthesis**

Living things grow. **Growth** increases the size from young offspring to mature adult

Living things must respond to what goes on around them. **Sensitivity** allows the body to detect and react to changes. The driver sees the lights change and responds by applying the brakes

Building a body

You were conceived by a sperm from your father joining with an egg from your mother. At that time you consisted of one cell! That cell would have had three main parts to it: a nucleus, cytoplasm and membrane – as the diagram shows.

That cell would have then divided to form two new cells (see Topic 4.6). These in turn also divided and grew repeatedly to produce the many trillions of cells that now make up your body.

As most new cells grow they also develop in different ways so that they can do different jobs. Their shape, size and the contents of the cytoplasm make them specialised to do different jobs well. The diagram opposite illustrates some of the ways cells specialise. Muscle cells can change shape and pull. Red blood cells do not have a nucleus, but are packed with a chemical called **haemoglobin** so they can carry oxygen. Nerve cells are long and thin; they have features that enable messages to pass along them quickly.

Cells that are similar to one another and do similar jobs are grouped together to form **tissues**. Muscle cells make muscle tissue, nerve cells make nerve tissue, and so on.

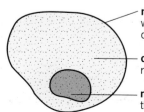

membrane – a thin layer that controls what substances pass into and out of the cell

cytoplasm – where most chemical reactions take place

nucleus – controls what happens in the cell

A typical animal cell

HUMANS AS ORGANISMS

Organs (e.g. **heart**, **lungs**, **brain**) contain several different types of tissue. Your heart, for example, is made of muscle tissue, nervous tissue and connective tissue. Organs cannot work by themselves; they depend on other organs to supply the things they need. They are linked together to form **organ systems**. The diagram shows some of the organ systems and the types of cells from which they are made.

The human machine

Your body is not just a collection of separate parts. Every system works with every other and each responds to changes in the other – like the parts of a complicated machine. When you run, for example, your skeleton acts like a set of levers, moving you onwards. But your muscular system must also pull on the right bones, in just the right way, to make them move. Your nervous system sends messages to coordinate the muscle movement. These muscles need a food supply (processed by your digestive system) and oxygen (brought in by your breathing system). Food and oxygen are transported to the muscles by your circulatory system, which then removes waste carbon dioxide.

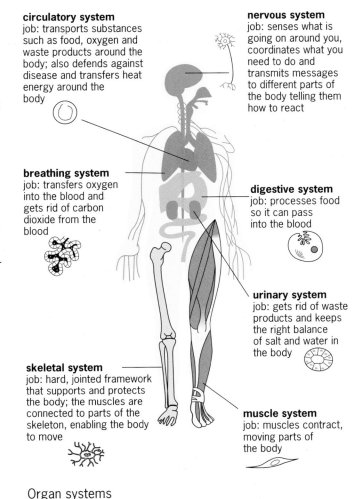

Organ systems

circulatory system
job: transports substances such as food, oxygen and waste products around the body; also defends against disease and transfers heat energy around the body

nervous system
job: senses what is going on around you, coordinates what you need to do and transmits messages to different parts of the body telling them how to react

breathing system
job: transfers oxygen into the blood and gets rid of carbon dioxide from the blood

digestive system
job: processes food so it can pass into the blood

urinary system
job: gets rid of waste products and keeps the right balance of salt and water in the body

skeletal system
job: hard, jointed framework that supports and protects the body; the muscles are connected to parts of the skeleton, enabling the body to move

muscle system
job: muscles contract, moving parts of the body

★ THINGS TO DO

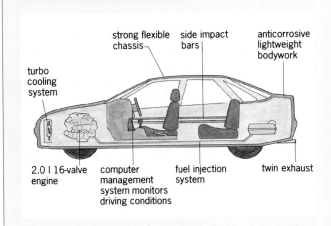

1 Look at the advert for a car.
   a) In what ways is a car similar to a living thing (e.g. a human)?
   b) Which systems of the human body do the same jobs as the parts shown?
   c) How is the human body more adaptable than a car?

2 A machine like a car can be rebuilt by replacing damaged or worn parts.
   a) Which parts of a human body can be successfully replaced (transplanted) if they stop working properly?
   b) Which parts of the body can be replaced by parts made by people?

3 **a)** Make a copy of the animal cell, the red blood cell and the nerve cell from the diagram above. Make a list of similarities and differences between each cell.

1.2 Muscles in action

Every muscle in your body has a job to do. Those in your upper arm control the movement of your lower arm. Those in your thighs control the movement of your lower leg. Your heart is one big muscle that pumps blood around the body.

Muscles

Muscles are organs that contract (i.e. get shorter and fatter) to force parts of your body to change shape or move. They can also relax (get longer and thinner). They are made of cells that can change shape easily. Because muscle cells need lots of energy they have many more **mitochondria** (where energy is released during respiration) than other cell types. A good blood supply provides the food and oxygen needed for respiration to take place.

The muscles that move the bones in your body are called **skeletal muscles** (see diagram). Other muscles change the shapes of organs. Food, for example, is squeezed along the gut by contracting and relaxing muscles. The heart is made of a special type of muscle, which can contract and relax constantly without tiring.

Regular exercise helps keep muscles fit. Combined with the right sort of diet, it also helps them grow. Some people exercise to highlight every external muscle on the body

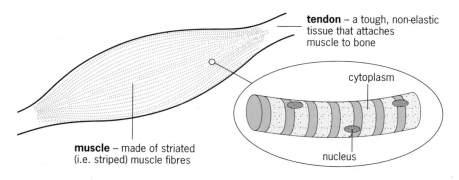

How skeletal muscle is made up

Muscles at work

The diagrams show what happens when you lift and lower a weight with your arm.

Notice that two muscles are involved each time; each has an opposite effect to the other. They work as an *antagonistic pair*.

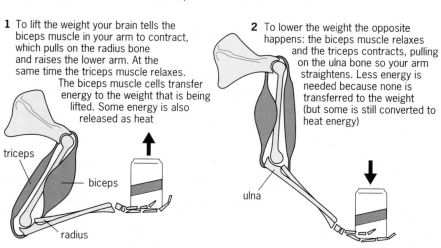

1 To lift the weight your brain tells the biceps muscle in your arm to contract, which pulls on the radius bone and raises the lower arm. At the same time the triceps muscle relaxes. The biceps muscle cells transfer energy to the weight that is being lifted. Some energy is also released as heat

2 To lower the weight the opposite happens: the biceps muscle relaxes and the triceps contracts, pulling on the ulna bone so your arm straightens. Less energy is needed because none is transferred to the weight (but some is still converted to heat energy)

Action of the biceps and triceps

HUMANS AS ORGANISMS

The bigger the better

This short article was found in a sports magazine.

Bigger muscles give more power, as they can work quicker. To develop more powerful muscles you need a good diet and regular training. The chart shows the correct ways to build up your muscles.

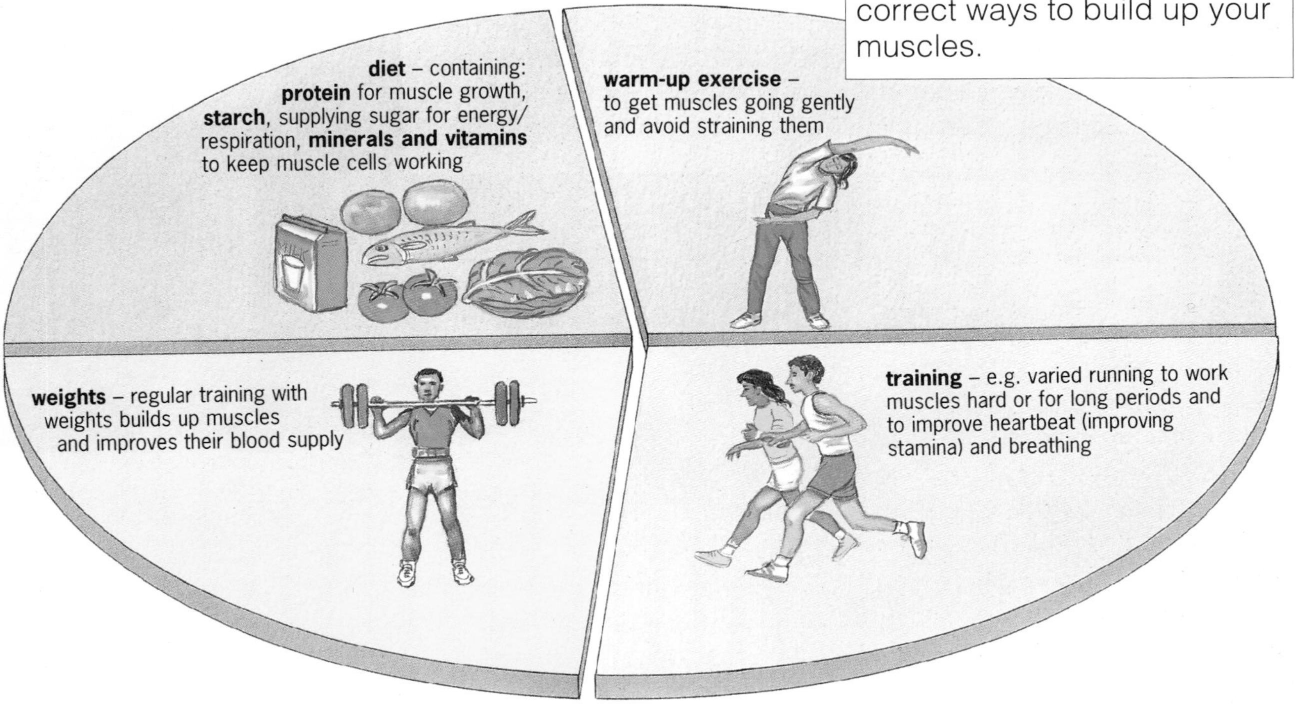

diet – containing: **protein** for muscle growth, **starch**, supplying sugar for energy/respiration, **minerals and vitamins** to keep muscle cells working

warm-up exercise – to get muscles going gently and avoid straining them

weights – regular training with weights builds up muscles and improves their blood supply

training – e.g. varied running to work muscles hard or for long periods and to improve heartbeat (improving stamina) and breathing

★ THINGS TO DO

1 Can people with bigger muscles lift heavier weights? Plan how you could test this idea. Make sure you think about things you must do to make sure your tests will be safe, and what you will need to do to make your test 'fair'. Carry out your tests after your teacher has checked your plan. In your account, include a section explaining your results.

RISK

2 What features of a muscle cell allow it to contract and relax easily? If possible, look at muscle cells from different parts of the body under a microscope or on a CD-ROM. Make a note of any similarities and differences between them. Try to explain any differences you notice.

3 **a)** When a broken bone is placed in plaster the muscles cannot move. Without regular exercise they begin to waste away – they get smaller. When the plaster is removed, the muscles must be built up again by regular exercise guided by a physiotherapist.
i) What exercise would you recommend for someone who had broken a bone in the lower arm?
ii) Why should they eat plenty of food containing protein?
iii) What kind of action should they avoid in the early stages?
b) Astronauts suffer from muscle wasting after spending months in space. What could have caused this? Design an exercise routine that they could follow in space to prevent muscle wastage.

1.3 A living framework

There are 206 bones in your body. Occasionally they are fractured or broken. Sometimes the ends of the bone are pushed apart, as shown in this X-ray photograph of a broken tibia (shin bone). If the person tried to stand up, the broken end of the bone would be pushed down into the fleshy tissue surrounding it, as it could not support their weight. The bone must be reset (the two ends must be brought together). The leg is then placed in plaster until new bone grows to heal the break.

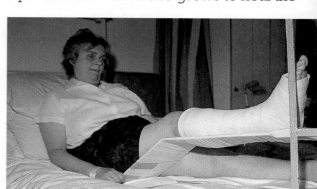

A patient with a broken leg

X-ray photograph of a broken (and displaced) tibia

The skeleton

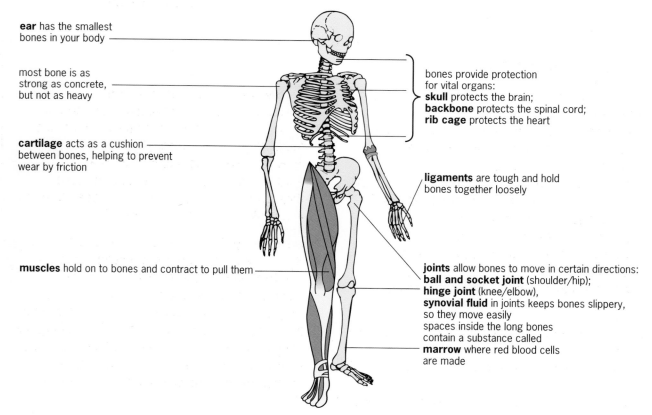

- **ear** has the smallest bones in your body
- most bone is as strong as concrete, but not as heavy
- **cartilage** acts as a cushion between bones, helping to prevent wear by friction
- **muscles** hold on to bones and contract to pull them
- bones provide protection for vital organs:
  skull protects the brain;
  backbone protects the spinal cord;
  rib cage protects the heart
- **ligaments** are tough and hold bones together loosely
- **joints** allow bones to move in certain directions:
  ball and socket joint (shoulder/hip);
  hinge joint (knee/elbow),
  synovial fluid in joints keeps bones slippery, so they move easily
  spaces inside the long bones contain a substance called
  marrow where red blood cells are made

The human skeleton

HUMANS AS ORGANISMS

The **skeleton** is a system of organs – **bones**, **cartilage** and **ligaments** – that work together. This skeletal system:

- supports us,
- protects other organs such as the brain, spinal cord, heart and lungs,
- has joints so that adjoining bones can swivel around one another,
- has places where muscles are joined to the bones so that we can move when the muscles contract. (A torn muscle means that it has torn away from the point where it is joined to the bone.)

Bone cells

Bone cells are alive (like most cells). Blood passes through tiny holes in the bone to supply them with food and oxygen and to remove the waste materials that they produce. It provides bone cells with minerals containing calcium and phosphate that they use to make calcium phosphate, which makes bone hard. They work throughout your life, adding or removing minerals that change the density and strength of your bones.

Children need a lot of these minerals in their food to help their bones develop. Vitamin D is also needed to help bone cells use minerals. A baby's bones are soft, almost rubbery. As you get older your bones become more brittle, and more easily broken (see Topic 1.21).

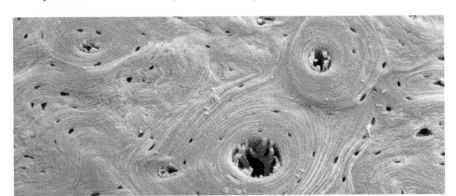

Bone cells; the large dark holes carry blood vessels

★ THINGS TO DO

1 Ask your teacher if you can examine some bones.
a) Leave a piece of bone in acid for a few days. The acid will dissolve salts in the bone. What happens to the bone?
b) Strongly heat a piece of bone for up to 20 minutes. The bone cells will burn. How is the bone different after heating?
c) Weigh different bones, measure their volume and then calculate their densities (density = mass/volume).
Why do you think some bones are more dense than others?

2 a) Imagine that on your way home from school your bones suddenly became rubbery.
i) Draw a picture to show what you would look like.
ii) Write a short story describing how you would move to get home.

b) Earthworms do not have bones.
i) Watch an earthworm and find out how it moves. Draw a series of pictures with notes describing what you think happens.
ii) Bones would be of little use to an earthworm; it would not be able to move through the earth as easily as it does. What do you think acts as their 'skeleton'?

3 Bones have different shapes and sizes. Some are hollow; others are solid. What do you think affects the strength of a bone? Make a note of your ideas, adding a reason for each one. Plan how you will test your ideas (you could use card to test them), including any safety precautions you need to take. Carry out your tests and write an account, including a section describing what your results mean.

1.4 A winning smile

Your teeth may still not gleam as much as the woman in the picture, but they are less likely to decay if you look after them. That means regular brushing, not eating too much sugar, avoiding eating between meals, and regular checks by a dentist. If you allow them to decay then you will suffer toothache. Teeth in really bad condition may even have to be removed!

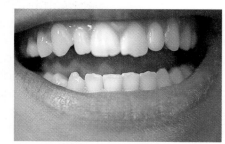

Good teeth need care

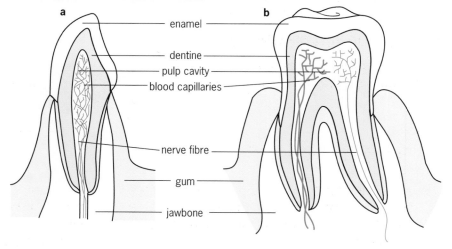

Section through
a an incisor tooth (at front of mouth);
b a molar tooth (at back of mouth)

What causes tooth decay?

Teeth are living structures. The cells are in the middle – the **pulp cavity**. They produce salts that make the outer, non-living layers (**dentine** and **enamel**) hard. Even though enamel is the hardest substance in your body, it cannot resist the corrosive effect of acids. Sugary foods easily stick to the surface of teeth and form a layer called plaque. In the warm, damp conditions inside your mouth, bacteria quickly change these foods into acid. Slowly the acid dissolves the enamel and dentine. When the pulp cavity is exposed to the air you have got a problem – toothache!

Preventing decay

Tooth decay can be reduced by:

- Regular brushing, which removes plaque. If the sugary plaque layer is removed the bacteria cannot produce acid.
- Eating sensibly. **Salivary glands** in your mouth produce **saliva**. Saliva is an alkali – it can neutralise acids. The pH in the mouth changes during the day as acid is produced and then neutralised by the saliva.

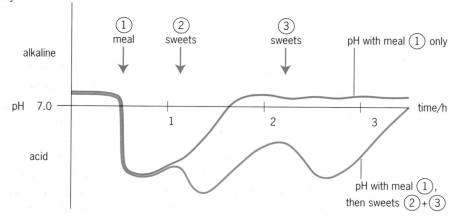

A graph of pH changes in the mouth

If you eat too often there is little time for saliva to neutralise acid in the mouth. So you should avoid eating between meals. Avoiding sugary foods or acidic liquids also helps.
- Taking fluoride. Fluoride is a mineral (chemical) that helps teeth to resist attack by acids. A slow, regular supply of fluoride is best and can be supplied in drinking water, in toothpaste or as tablets.
- Visiting the dentist. The dentist can quickly spot signs of decay and deal with it.

Why do we have different teeth?

You have different types of teeth because they must do different jobs as shown in the diagram.

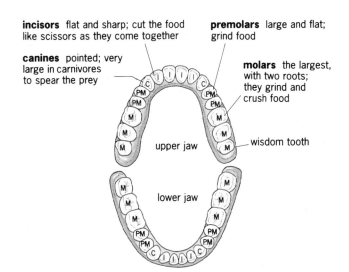

The arrangement of teeth in the upper and lower human jaw (adult)

★ THINGS TO DO

1 Use a mirror to look at your own teeth. Copy the diagram of the teeth in the upper and lower jaw. Mark on it any teeth that are missing, or have been filled.

2 Disclosing tablets when chewed stain any plaque on teeth.

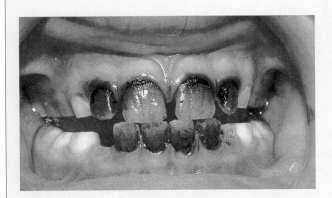

How can they be used to help prevent tooth decay? If possible, try chewing a disclosing tablet yourself. You could use it to see how well brushing removes plaque.

3 The teeth of a **herbivore** (a plant-eating animal) such as a sheep are designed to wear away. This exposes the dentine, which wears away quicker than the outer enamel. Sharp ridges are left – similar to those on a metal file – which are ideal for grinding the hard grasses that sheep eat.

Look at the photo of a sheep's molar teeth.

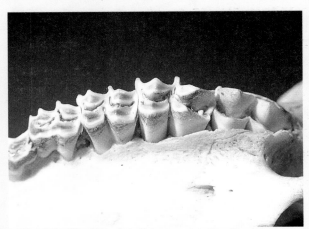

a) Why is it important for the teeth of sheep to grow continuously?
b) The sheep's jaws move over each other in a circular movement as food is chewed. How will this help?
c) Why do sheep not need canine teeth?
d) If possible, compare these teeth with those in a jaw of a carnivore (e.g. a dog). What are the similarities and differences?

1.5 Food is important

Like all other animals, you need food to survive. Food provides chemicals and energy to keep your body working and healthy. The food you eat – your **diet** – must be both *adequate*, to give you all the energy you need in a day, and *balanced*, to provide the right amount of each of the chemicals your body needs.

Preparing for the race

GREAT, I'VE BEEN ACCEPTED FOR THE MARATHON. THAT GIVES ME THREE MONTHS TO PREPARE.

SHOW ME WHAT IT SAYS.

Advice on what to eat
Keep your diet balanced by eating these foods:
carbohydrate (sugar or starch) for energy,
protein for growth and tissue repair,
fat to store energy and make cell membranes,
minerals for important body chemicals,
vitamins A, B, C and D for cells to work properly,
fibre for a healthy digestive system,
water to make up for daily losses.
Remember – you must eat enough food to replace the energy you need each day. Keep your diet adequate.

This newspaper article is about the problems many athletes have with their diets.

Athletes at risk
These graphs show the range of body weight for adult men (right) and women of different heights. Any weight inside the green area is 'normal'.

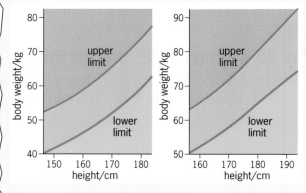

Nearly half of the leading distance runners suffer from 'eating disorders'. A study of 200 athletes who compete over 800 m showed that the runners had poorer diets than a control group of 200 non-runners of the same age, size and sex. Although the non-athletes had slightly better diets, they tended to eat too much. They were more likely to be overweight because the extra food (over and above what they really need each day) was stored in their body as fat. The athletes ate very little food and often missed out fatty foods entirely to keep their weight down. Some had symptoms of anorexia nervosa: they were very thin and lacked the energy and chemicals to stay healthy.

You need to know how much of each kind of food you require at different ages to keep your diet balanced and adequate, so you stay healthy. These details about food and diet requirements can help you understand your needs.

Data adapted from *Food Tables*, AE Bender and DA Bender, Oxford University Press 1986

Diet table

	Amounts needed per day by people of different age and gender				
	15–17 years		18–34 years		
gender	male	female	male	female	female (pregnant)
body weight/kg	60	58	72	60	63
energy/MJ	12.0	9.0	12.0	9.5	10.0
protein/g	72	53	72	54	60
calcium/mg	600	600	500	500	1200
iron/mg	12	12	10	12	13
vitamin A/μg	750	750	750	750	750
vitamin C/mg	30	30	30	30	60

HUMANS AS ORGANISMS

Food chemicals

Name of chemical	Good sources	Problems suffered if too little is eaten (deficiency symptoms)
carbohydrate – sugar – starch	fruit, jam bread, potatoes, pasta	lack of energy
protein	meat, fish, eggs, beans, seeds, grains	stunted growth, thin muscles
fat	fatty meat, butter, milk, oils, oily fish, nuts, avocados	lower immunity, heart and circulation problems ('unsaturated' fats only)
minerals – calcium – phosphate – iron	cheese, milk, canned fish, green leafy vegetables, figs cheese, milk, meat, fish greens, liver, shellfish, red meat, soy beans, eggs	weak bones, high blood pressure, PMS weak bones, lack of energy anaemia, poor mental function
vitamins – A – B – C – D	liver, eggs, milk, carrots and other orange vegetables (as beta carotene) wholegrains, eggs, liver, green leafy vegetables sweet peppers, acid fruits, sprouts, tomatoes, broccoli, cabbage oily fish, milk, eggs	stroke, blood vessel damage, poor eyesight, lower immunity, poor skin bad memory and concentration, liver damage, birth defects scurvy, gum bleeding, asthma/bronchitis, heart and circulatory problems, high blood pressure weak bones (rickets), cancer (esp. breast)
fibre	whole cereals, bran, oats, vegetables, fruit	digestive problems, gallstones, kidney stones, colon cancer

★ THINGS TO DO

1 The weights and heights of three adult men are:
   Lewis   1.75 m, 62 kg
   Ahmed   1.65 m, 67 kg
   Rik   1.85 m, 112 kg
   Runners like to keep their weight at the lower limit. Javelin throwers prefer a weight near to the upper limit. Use the body weight graph to decide:
   a) who is most likely to be a runner,
   b) who is most likely to be a javelin thrower,
   c) who is unlikely to be an athlete.

2 The amount of energy someone needs depends what they do. Being more active means that more energy is needed. A male road worker, for example, needs 20 000 kJ of energy each day, whereas a male computer operator (of the same size) needs only 11 500 kJ. You need more energy when you run (45 kJ min^{-1}) than when you walk (16 kJ min^{-1}). Look at the diet table opposite and explain how the amount of energy needed varies with:
   a) size,
   b) gender (female or male).

3 You can find out what is in food by looking at the nutrition details on food packets, in data books or on a computer database.
   a) List everything you eat in 24 hours.
   b) Work out what these foods contain.
   c) Compare this with what is recommended in the diet table for someone of your age.
   d) The diet table can be considered only as a rough estimate of what you need. What other factors could influence your daily needs?

1.6 It's your choice

Your diet may relate to your lifestyle, your background, or what you believe in (e.g. your religious faith). Some people prefer not to eat certain types of meat, for example. Others will not eat any foods that come from animals. As long as the diet you choose is both adequate and balanced you will stay active and healthy.

Different types of diet

A mixed diet or not?

Our teeth and digestive system are designed to process all sorts of food – soft meat from animals and harder foods from plants. This is why we could be called **omnivores**. Most people eat a mixture of food from animals and plants, but many people are vegetarians or even vegans – they eat only plant foods.

Animals like sheep and cows can digest plant food better than we can. These plant-eating animals (herbivores) can produce an **enzyme** called **cellulase**, which digests plant cell walls so further digestion is possible. Humans lack this enzyme.

Plant foods tend to be poorer sources of protein, iron and vitamin B_{12}, but a carefully chosen mixture of plant foods gives vegetarians all they need. Studies suggest that vegetarians have lower risks of heart disease and cancer than do meat eaters.

Wholewheat bread or not?

Wholewheat bread contains a lot of fibre (roughage). Bread made with refined flour lacks fibre, as refining removes it. Fibre is a mixture of substances, mostly lignin and cellulose from plant cell walls. Fibre is not digested, so passes through the digestive system (**gut**) and is released from the **anus** as waste. But, if you're thinking that fibre is useless, think again. It makes it easier for food to move along the gut (see below) and reduces the risk of bowel cancer. Getting a taste for wholewheat bread and other foods high in fibre may keep you healthier.

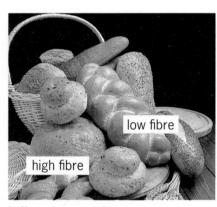

Different types of bread

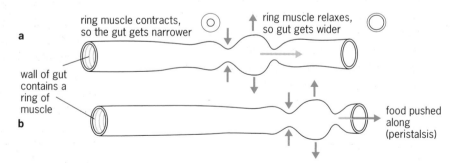

Food is squeezed along the gut by muscles in the gut wall as they contract and relax. **a** This process is called **peristalsis**. It is less effective when food is digested and dissolves. Muscles can push against undigested solid food more easily. Sometimes food is held in places, to allow time for it to be digested and absorbed. Food is trapped inside the stomach when ring muscles at each end of this organ tighten. Peristalsis then mixes food around, like clothes being tumbled in a washing machine. **b** When the food is ready to move on, the bottom ring muscle relaxes and food is pushed out into the small intestine

How good is milk?

Milk is a 'natural food' made by mothers to feed their young, although some (human) mothers prefer to give their babies cow's milk. Milk is the only food that very young babies can have and contains everything that the baby needs, except vitamin D. The baby makes this vitamin when its skin is exposed to sunlight. Perhaps surprisingly, adults cannot digest milk as well as babies can – many adults in the world become quite sick when they drink milk.

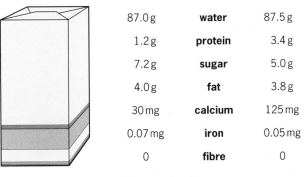

human milk		cow's milk (complete)
87.0 g	water	87.5 g
1.2 g	protein	3.4 g
7.2 g	sugar	5.0 g
4.0 g	fat	3.8 g
30 mg	calcium	125 mg
0.07 mg	iron	0.05 mg
0	fibre	0

The components of human and cow's milk (mass per 100g)

★ THINGS TO DO

1 a) Explain the difference between semi-vegetarians, lacto-vegetarians and vegans.
b) Do a survey to find out how many people out of 20 are vegetarians or eat vegetarian meals. Pool your results to compare different groups of people (e.g. by age or gender).

2 a) Show the information for each type of milk shown in the illustration as a bar chart.
b) Describe the differences between human milk, cow's milk and the dried milk used for babies. Are the differences likely to have any real effect on a baby's growth?
c) Give three reasons why adults need less milk in their diet than babies do.

3 Cooking can take the goodness out of food. The table shows some tests on potatoes to see how much vitamin C they contain.
a) Which potatoes have the most vitamin C?
b) Which treatment seems to remove most vitamin C?
c) Look back to Topic 1.5 to find out why vitamin C is important for you.
d) What health advice would you give to someone who likes eating boiled potatoes?
e) Many people like to eat potatoes as chips. How might cooking potatoes in this way affect the vitamin C content? Find out how to test for vitamin C and write a detailed plan describing how you could test your idea (called your **hypothesis**).

Type of treatment	Amount of vitamin C /mg cm^{-3}
raw potatoes	15.0
put in cold water, boiled and simmered for 15 min	10.0
put in hot water, boiled and simmered for 15 min	8.5
put in hot water, boiled and simmered for 30 min	7.5

1.7 Processing food

What is food?

We think of food as what we eat. But it may come in a number of shapes and forms, as shown in the photographs.

Foods such as these contain hundreds of different chemicals; it is these chemicals that the body needs

Other forms of food look quite different. Astronauts, for example, cannot prepare 'normal' food in space. Their food is a concentrated liquid that is squeezed from tubes

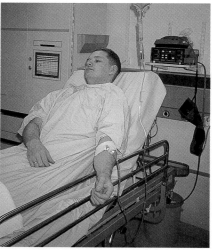

Some hospital patients may not be able to eat food at all. They are fed intravenously, and the food goes directly into the blood. The food here is in solution

The important chemicals in the food you eat, which supply the energy, minerals and nutrients you need, are carried to every cell in your body by blood. But first they have to get into your blood.

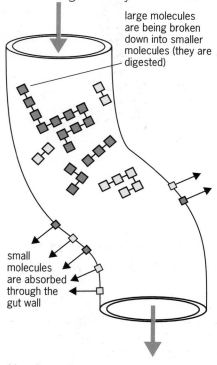

How large food molecules are broken down and absorbed

Most solid food must be changed before this can happen – it has to be digested. **Digestion** is a process that chemically breaks down large, insoluble molecules in food into smaller, soluble molecules. When they are small enough they can dissolve and pass through the gut wall into the blood; this process is called **absorption**.

- Carbohydrates like starch are digested into **sugars** (e.g. **glucose**).
- Proteins are digested into **amino acids**.
- Fats are digested into **fatty acids** and **glycerol**.

Chemical breakdown

Digestion is speeded up (catalysed) by chemicals called enzymes. Different types of enzymes are made by the organs of the digestive system.

Where is food absorbed?

1. Some small molecules such as sugar, water and minerals are absorbed into blood **capillaries** through the stomach wall.
2. Most other molecules are absorbed through the small intestine wall, including sugars, water, amino acids, glycerol, minerals and vitamins (note that the sugars, water and minerals absorbed here are produced by digestion of food *after* it has passed from the stomach).
3. Fatty acids are absorbed through the small intestine wall into a liquid called **lymph**, which eventually flows into blood vessels.
4. The surface area of the small intestine wall is very large, because it has lots of tiny projections called **villi**.

HUMANS AS ORGANISMS

Food is processed (digested and absorbed) as it passes along your digestive system (**gut**). The gut is a long tube made of organs that include the mouth, stomach and intestines

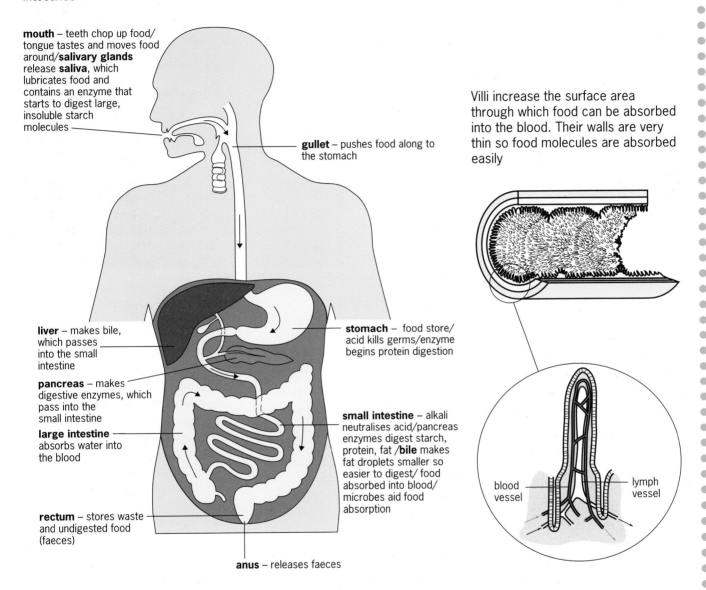

mouth – teeth chop up food/ tongue tastes and moves food around/**salivary glands** release **saliva**, which lubricates food and contains an enzyme that starts to digest large, insoluble starch molecules

gullet – pushes food along to the stomach

Villi increase the surface area through which food can be absorbed into the blood. Their walls are very thin so food molecules are absorbed easily

liver – makes bile, which passes into the small intestine

pancreas – makes digestive enzymes, which pass into the small intestine

large intestine absorbs water into the blood

rectum – stores waste and undigested food (faeces)

stomach – food store/ acid kills germs/enzyme begins protein digestion

small intestine – alkali neutralises acid/pancreas enzymes digest starch, protein, fat /**bile** makes fat droplets smaller so easier to digest/ food absorbed into blood/ microbes aid food absorption

anus – releases faeces

blood vessel

lymph vessel

★ THINGS TO DO

1 Amylase is an enzyme produced by the salivary glands in your mouth. An investigation was done to test the idea that amylase digests starch. It was found that starch disappeared from the tube containing amylase, but if there was no amylase the starch was not affected. You might conclude that starch is digested by amylase.

But from this investigation you could *not* prove that starch had been changed into sugar. Why not?

Write down a plan of how you could show that sugar was made. If possible carry it out. You might also find out how quickly the starch disappears when it is mixed with amylase.

1.8 Enzymes at work

Food doesn't always go where it should! Food stains are difficult to remove from clothing because they contain large, complicated molecules that are usually insoluble (they do not dissolve). Biological washing powders contain enzymes. The enzymes in your digestive system speed up the breakdown of the large food molecules into smaller molecules. They do the same job as those in washing powders. The smaller, soluble molecules dissolve in the wash, leaving the clothes stain free.

The most difficult stains to remove are from protein foods like egg or spills of blood. Protein-digesting enzymes (proteases) remove these stains.

The enzymes work best at temperatures higher than body heat (37 °C) – the extra heat helps the detergent in washing powder to disolve stains non-biologically, but excessive overheating (i.e. boil washing) prevents a biological powder working well as it destroys the enzymes

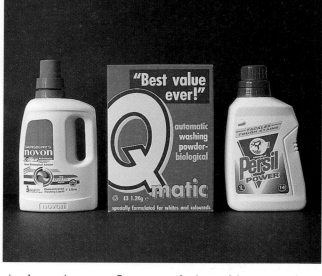

Enzymes are an important group of chemicals that speed up chemical reactions in your body. They act as catalysts. An enzyme is unchanged at the end of the reaction that it speeds up.

Enzymes are called **biological catalysts** and have the following properties:

- they are made of protein,
- they are specific, speeding up one reaction only,
- they work best within a small pH range, which is different for each enzyme,
- they work fastest when warm, usually at body temperature (37 °C),
- above 50 °C their shape is changed (they are *denatured*) and they stop working.

There are many different enzymes in your body, speeding up the different reactions that keep you alive. The enzymes are made in the cells, and work mostly in the cytoplasm. Some speed up reactions where large molecules are broken down into smaller ones. Others do the opposite – they speed up the **synthesis** (building up, also called **assimilation**) of large molecules from smaller ones.

An example is making protein from amino acids; this happens in growing cells and when new enzymes are being made. Another type of enzyme helps energy transfer from food to the molecule **adenosine triphosphate** (**ATP**) during respiration. This molecule is used by the body to store energy released from food breakdown and other sources.

Your digestive system makes the following digestive enzymes (see table below):

Type of enzyme	Where made	Where it works	Reaction speeded up
carbohydrase	salivary glands pancreas intestine wall	mouth small intestine small intestine	starch to sugars
protease	stomach pancreas intestine wall	stomach small intestine small intestine	protein to amino acids
lipase	stomach (very little) pancreas small intestine wall	stomach small intestine small intestine	fats to fatty acids + glycerol

Shape is important

Each enzyme works on some types of food molecule but not on others. This is because enzymes will only fit on to food molecules that have a complementary shape.

Tubes with egg white suspension plus pepsin, becoming clear as egg white protein is digested

1 The shape of the enzyme lets it fit on to part of the food molecule
2 Enzyme and food molecule are held together as a 'lock and key'
3 When the enzyme flexes slightly the food molecule is pulled apart to form two new substances
4 The two new food molecules separate
5 The process may be repeated until the large molecule can be broken down no further

The 'lock and key theory' explains how the enzyme helps the molecule to break down

Digestion in action

Pepsin is a protease enzyme found in the stomach. It digests protein.

Egg white is mostly protein. Pepsin will digest the protein in solidified egg white. This then dissolves, so you cannot see it. In the stomach, food is mixed with this enzyme for up to 6 hours, so that it has time to work.

Pepsin works best in acidic conditions. It stops working when it passes, with the food, into the small intestine. The conditions here are slightly alkaline and favour a different set of enzymes. Any acid passing from the stomach is neutralised by a solution including bile. Bile is made by the liver and stored in the **gall bladder**. Bile also emulsifies fats, breaking down large fat drops into smaller ones. This increases the surface area so that lipase enzymes can digest the fat quicker.

★ THINGS TO DO

1 The graph shows how the activities of pepsin and amylase (a carbohydrase) vary with pH.

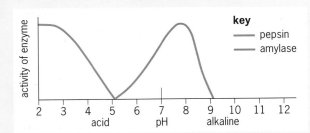

a) Copy the diagram into your book.
b) At which pH does pepsin work best?
c) Pepsin works in the stomach. Look back to Topic 1.7 to find out how the stomach keeps its pH just right for pepsin to work at its best.
d) At which pH does amylase work best?
e) Amylase is released into the small intestine. Food passing into here from the stomach has a pH of about 3. What must happen to it before amylase starts working on it?
f) Why will any pepsin passing into the small intestine stop working?
g) Amylase is also released into your mouth in saliva. Why is it helpful that saliva also contains an alkali?

2 a) Apart from pH, what factors could affect how well pepsin works?
b) How could you use egg white and pepsin to test your ideas? Have your ideas checked before starting any practical work.
c) Write a detailed plan of your investigation and findings.

3 Think about the lock and key model that explains how an enzyme works. Try to draw diagrams to explain why a denatured enzyme does not work.

1.9 Breath of life

A kiss saves Steve
by our local reporter

The life of a gardener was saved by a kiss yesterday. Steve Redrose was working in his garden when he was stung by a wasp. Within a few minutes he was gasping for breath as his nervous system was affected by the venom.

Fortunately for Steve, Jenny Agate, a 19-year-old trainee nurse, was passing by on her way home. She told us 'I noticed Steve lying on the ground and ran across. He had stopped breathing. I began artificial ventilation immediately while his neighbour phoned for an ambulance.' A hospital spokesperson said 'this man would have died within a few minutes if Jenny had not acted so quickly. The nervous system controls breathing. When something interferes with the nervous system it can have all sorts of effects. In this case the breathing system was affected by the sting – because he was allergic to the substances in it. Brain cells begin to die after 3 minutes without oxygen. Steve owes his life to the young lady.'

The energy released from sugar is used to:
- allow muscles to contract,
- make large molecules from smaller ones,
- keep the body temperature constant,
- pump certain substances into and out of cells.

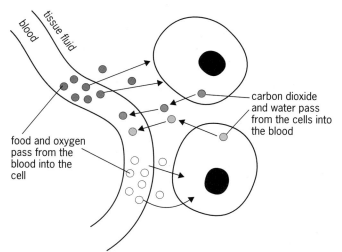

Exchanges between a blood vessel and cells

Breathing for respiration

Breathing is vital to take in oxygen needed by cells to respire. Cells in your body release energy by respiration when sugars react with oxygen. Sugars are absorbed through the wall of the gut into the bloodstream, and carried to the cells (see Topic 1.7). Oxygen (from the air that you breathe in) is absorbed into the blood in the lungs. The blood therefore carries oxygen *and* food to the cells.

Respiration can be summarised as:

sugar + oxygen → carbon dioxide + water + energy
$C_6H_{12}O_6 + 6O_2 \rightarrow 6CO_2 + 6H_2O$ + energy

The energy is stored in **ATP** (see Topic 1.8). The carbon dioxide and water are waste products. They pass into the blood, which carries them to the lungs and out with the air you breathe out. The full name for this process, which uses oxygen, is **aerobic respiration**.

Supersprint

A sprinter in action

In races such as that shown in the photograph, the cells in the leg muscles of runners may release energy without oxygen as they dash for the finish, or if they push themselves too hard. Their breathing and circulation systems cannot supply oxygen fast enough.

The energy has been released from food by **anaerobic respiration** (respiration without oxygen). During anaerobic respiration less energy is released from food and a waste chemical called lactic acid is made (instead of the carbon dioxide produced during aerobic respiration).

During anaerobic respiration:

$$\text{sugar (glucose)} \rightarrow \text{lactic acid} + \text{energy (stored as ATP)}$$

Your own body sometimes releases energy by anaerobic respiration for a short time when you exercise vigorously. If you have a cramp when running hard, this is a sign for your body to slow down. A cramp (muscle fatigue) occurs when too much lactic acid collects in the muscles. It disappears as the lactic acid dissolves in the liquid part of the blood and is taken to the liver. It is then converted into carbon dioxide (and removed through the lungs) and a little more energy is released.

In the liver:

$$\text{lactic acid} + \text{oxygen} \rightarrow \text{carbon dioxide} + \text{water} + \text{energy (stored as ATP)}$$

Notice that oxygen is needed here. The lactic acid that builds up in your body as anaerobic respiration takes place creates a need for oxygen – an 'oxygen debt'. After vigorous exercise you continue to breathe hard to take in the oxygen you need to pay it back.

★ THINGS TO DO

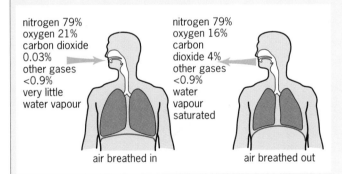

air breathed in — nitrogen 79%, oxygen 21%, carbon dioxide 0.03%, other gases <0.9%, very little water vapour

air breathed out — nitrogen 79%, oxygen 16%, carbon dioxide 4%, other gases <0.9%, water vapour saturated

1 The illustration shows the differences between air that you breathe in and air you breathe out.
   a) Draw two pie charts to illustrate this data.
   b) Explain the differences in the figures for oxygen.
   c) Explain the differences in the figures for carbon dioxide.
   d) How could you demonstrate that the air you breathe out is saturated with water vapour?

2 The apparatus shown can be used to demonstrate that you breathe out more carbon dioxide than you breathe in. How could you use it to test for differences between the carbon dioxide that different people breathe out?

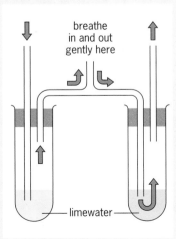

1.10 Breathing all your life

At ground level there is enough oxygen in the air to supply your body's needs. At high altitudes, as in the photograph, there is less oxygen in the air.

Breathing

Breathing does two things:

- it gets the oxygen needed by all cells into the blood,
- it gets rid of the carbon dioxide produced by cells from the blood.

When you breathe *in*, air is not 'sucked in' – it is pushed in by atmospheric pressure (see *GCSE Science Double Award Physics*, Topic 3.5), which is higher than the pressure inside your lungs. Your lungs are like sponges that fill with air. The organs of the breathing system include the lungs and also the throat (pharynx), windpipe (trachea) and bronchi, the diaphragm and associated muscles. The diagrams show how we breathe.

Mountaineers climbing the world's highest peaks get the extra oxygen they need by breathing oxygen from gas bottles. Without the extra oxygen they can move only 10 or 20 steps before stopping for rest; there is not enough oxygen in their blood to release the energy their muscles need to keep moving

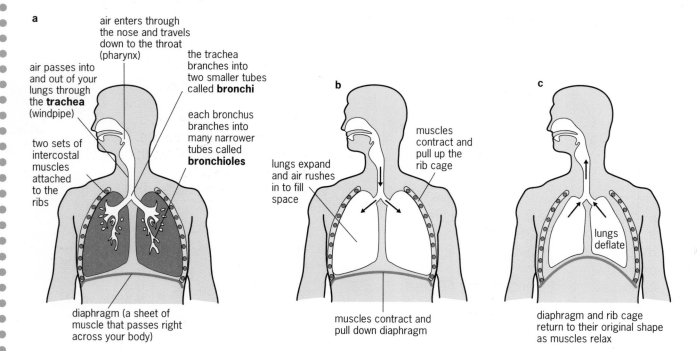

a Organs of the breathing system; and movements of the chest: **b** inhaling; **c** exhaling

HUMANS AS ORGANISMS

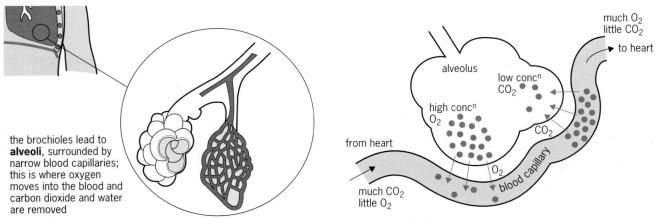

the brochioles lead to **alveoli**, surrounded by narrow blood capillaries; this is where oxygen moves into the blood and carbon dioxide and water are removed

a Detail of alveoli with surrounding capillaries; **b** O_2 and CO_2 movements in the lung

Diffusion

Diffusion is the movement of particles from a place where they are common (high concentration) to a place where there are few (low concentration). The gases oxygen and carbon dioxide move into and out of each alveolus by diffusion.

Diffusion is fairly slow. To make up for this the surface area of the lungs is as large as possible, with millions of alveoli. Diffusion takes place when there is a big difference in the concentrations. Breathing in and out creates this big difference, so the oxygen and carbon dioxide diffuse rapidly.

Oxygen also diffuses into cells from the blood in capillaries, and carbon dioxide diffuses from cells to blood.

More and more

The amount of air breathed in depends on:

- the rate of breathing – how many times you breathe in and out per minute,
- the depth of breathing – how much air passes into the lungs each time you breathe in.

★ THINGS TO DO

1 Feel your chest move up and out when you breathe in.
  a) What happens to your chest when you breathe out?
  b) Use a tape measure to calculate the change in chest size as you breathe.
  c) What is the maximum difference in chest size you can create?
  d) You can measure your maximum lung capacity by using the apparatus shown below.
  You breathe in fully, then blow out fully into the container of water. The amount of water pushed out as you breathe out is your maximum lung capacity.
  i) What is your maximum lung capacity?
  ii) How does it compare with the measurements of other people?
  iii) What factors could affect the size of someone's lung capacity?
  iv) How could you test your ideas?

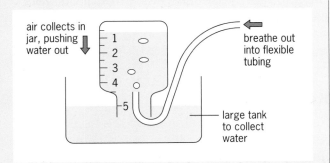

1.11 Hale and hearty

World class sportspeople such as Jonathon Edwards need to be fit. They train several times each week. Training may include running, swimming, or working with weights (depending on the sport)

Other people also need to stay fit. Young people need to be fit so that they can cope well with the variety of activities they will do – playing, thinking, writing, running and so on

Although it may seem that the world champion in the photograph is fitter than the young child, this is not necessarily so. Fitness simply means that your body is ready to cope with the everyday activities that your lifestyle demands. Different people may show different types of fitness:

- A weightlifter needs strength.
- A gymnast needs a supple body.
- Long-distance runners need stamina.
- Sprinters need speed.

A fit person will have all of these to some degree. One thing which they all have in common, however, is a healthy heart. The heart is a large muscle that works throughout your life, pumping blood around your body. It must keep blood flowing 24 hours each day, 365 days of each year, throughout your life. To do this it beats about 37 million times each year: it needs to be fit to do that! You must take steps to make sure that it stays in good condition.

Keep it healthy

Regular exercise helps keep the heart muscle in good condition, in the same way that weightlifting builds up the muscles of the arms. The heart can then pump more blood at each beat, so it needn't work as hard. Combined with a sensible diet, exercise can also help to keep the **arteries** and **veins** clear (see Topic 1.13), so blood flows easily through them.

If you eat a lot of fatty foods, the fat can stick to the inside of the arteries, partially blocking them. As their diameter is reduced, the blood flow through them is also reduced and the heart has to work harder to maintain the blood supply that you need.

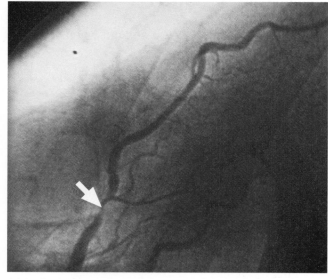

A blocked blood vessel (the arrow shows the site of blockage)

HUMANS AS ORGANISMS

In Britain someone suffers from a fatal heart attack every 3 minutes. Heart disease is the biggest killer in the UK. During an attack one of the coronary (i.e. heart) blood vessels may become partially blocked. This reduces the supply of oxygen to the heart muscle, causing angina, a painful cramp. If the person rests immediately, the muscle recovers. A serious blockage, however, may completely stop the blood flow to part of the heart muscle, resulting in a heart attack.

The lack of oxygen damages the heart muscle. Some of the cells die, and are replaced by scar tissue, which prevents that part of the muscle working properly. If the area affected is small, the person normally recovers. However, if the damaged muscle disturbs the regular beat of the heart, the heart may stop (cardiac arrest) and the heart attack may prove fatal.

★ THINGS TO DO

1 a) Collect leaflets from local health centres showing the commonest causes of heart disease.
b) Use the information to design a poster to illustrate the problems of heart disease.

2 The government has set 'Health of the Nation' targets for the coming years. One is to reduce the level of ill-health and death caused by coronary heart disease, and the risk factors associated with it.

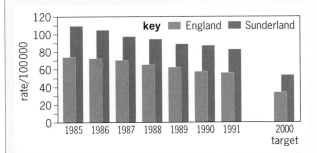

Death rates (under 65 years) from coronary heart disease nationally, and in Sunderland (from *The Health of Sunderland*, annual report 1993, commissioned by the Sunderland Health Authority, p. 16, with kind permission)

a) What differences do you notice between the pattern of heart disease in Sunderland and the national figures?
b) What might be the reasons for such a difference?
c) What steps could national and local authorities take to reach the targets set by the government?

d) Imagine you work for a local authority. You must decide what steps the local authority can take to meet government targets. Make a list of things that could be tackled by the authority; describe some ways in which the message can be got across to the people.

3 The table shows the total size of the heart of some trained athletes.

Athlete	Heart volume/cm³
racing cyclist	1100
canoeist	1100
long-distance runner	950
footballer	900
sprinter	800
ice hockey player	750
weightlifter	700
average man	700

a) Plot the data as a bar graph.
b) Why do athletes in some sports have bigger hearts than others?
c) From this evidence, what sort of exercise makes the heart grow most? What other information would be needed to be sure that your answer is correct?
d) All of the information in the table is for male athletes. What differences would there have been if the table had also shown female athletes?

1.12 Living pump

Your heart works in a similar way to this water pistol.

A living pump

Your heart is continuously squeezing in and relaxing, pumping blood through the chambers and around your body. It works as two pumps, side by side, which operate together. The flow of blood through the heart is controlled by valves between the chambers. The right side pumps deoxygenated blood (i.e. which has already transferred its oxygen to the cells) to the lungs, where it collects more oxygen. The left side pumps oxygenated blood (i.e. with fresh oxygen in it) to the muscles and other organs of the body. There is a two-way journey through the heart for the blood; therefore it goes through the heart twice for every once around the body.

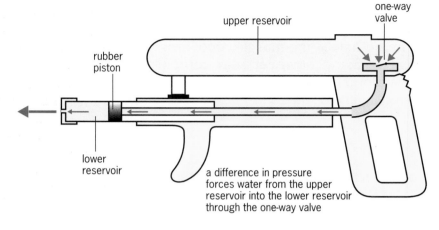

How a 'superpistol' works

Heartbeat

Like any other pump, the heart has to fill up before it can empty. The heartbeat is felt at the emptying stage as both pumps contract to push blood out. The sounds you can hear are the valves in the heart snapping shut after blood has been forced through.

Heart rate

The number of times your heart beats each second is your heart rate. An adult's average heart rate (when resting) is about 70 beats per minute; a newborn baby's is much higher. The rate varies depending on what you are doing, and how fit you are. For example, if you are doing strenuous exercise, your muscle cells must release energy rapidly. Food and oxygen must be supplied more quickly, so your heart (and lungs) must work faster.

You can feel a heartbeat, or pulse, in your wrist where an artery passes over one of the bones. Each pulse is a squirt of blood that has been pushed along by the heart.

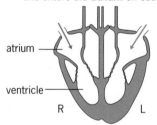

1 Blood flows to the heart through the **veins** and enters the **atrium** on each side

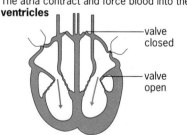

2 The atria contract and force blood into the **ventricles**

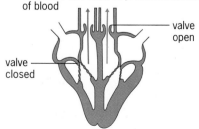

3 The ventricles contract to push blood out of the heart; also **valves** close between the ventricles and atria, to prevent backflow of blood

How the heart pumps blood

The busiest muscle

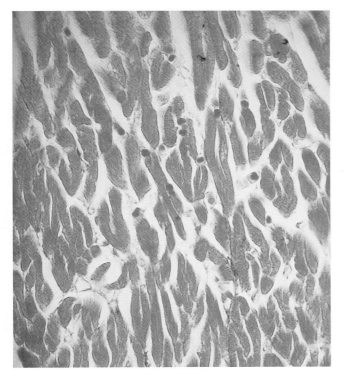

Like other muscles in the body, the heart needs its own supply of oxygen and food. So it needs its own blood supply. Oxygen is provided by the pulmonary vein (which brings freshly oxygenated blood from the lungs to the left atrium). Food is brought by the coronary vein (the 'vena cava', which returns blood from the body back to the right atrium).

Valves inside the heart ensure that blood flows one way only. Diseased or torn valves reduce the effectiveness of the heart as a pump because blood can flow back in the opposite direction. Faulty valves can now be replaced with either artificial valves or valves taken from animal hearts.

The heart contains cardiac muscle tissue, which, unlike other muscle, can contract and relax regularly throughout your life without tiring. The cells are branching muscle fibres, which contain a lot of mitochondria (see Topic 1.2) to release the continuous supply of energy the heart needs

★ THINGS TO DO

1 As two men raced against each other over a mile their pulse rate was measured. The graph shows the results.

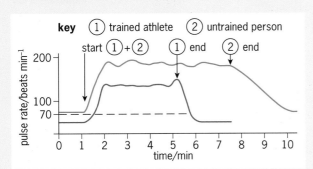

a) What happened to pulse rates at the start of the run?
b) What was the maximum heart rate for the untrained runner?
c) What was the maximum heart rate for the trained athlete?
d) Why does the heart have to beat faster when you exercise?
e) How could the trained athlete manage to run the race with a slower heart rate than the other?
f) What happened after the run? Compare the trained athlete with the non-athlete. Try to explain any differences you notice.
g) Lack of exercise can cause veins and arteries to fur up, reducing their diameter. How will this affect their activity?

2 The US Navy measure fitness by measuring someone's heart rate, getting them to do a certain exercise, then timing how long it takes for their heart rate to return to normal. The longer it takes, the less fit the person is. With others in your group write a plan to find out whether someone's fitness would improve with regular exercise each day for 1 month. You may be able to find a volunteer who would be willing to try this on behalf of the class.

1.13 Blood networks

The circulatory system

The blood vessels, the heart and the blood itself make up the circulatory system. A complicated network of blood vessels (arteries, veins and capillaries) carry blood around your body. The blood, pumped around by the heart, must flow in the right direction to do its job effectively. The circulatory system is rather like a one-way road system – it makes sure that everything flows in the right direction. Blood is pumped through the heart twice for every once it goes round the body – this is called a 'double circulation system'.

The blood carries the oxygen and food to the cells. These release the energy you need. Energy is needed by your body throughout each day, so the cells must receive a continuous supply of food and oxygen. Every cell in your body needs substances that keep it alive and allow it to do its job effectively. Cells also make waste substances, such as carbon dioxide and water, which must be removed quickly (carbon dioxide is a poison in excessive amounts). The circulatory system therefore acts like a collection and delivery service, picking up substances where they are plentiful and dropping them off where they are needed. It also acts as a central heating system, spreading heat around the body.

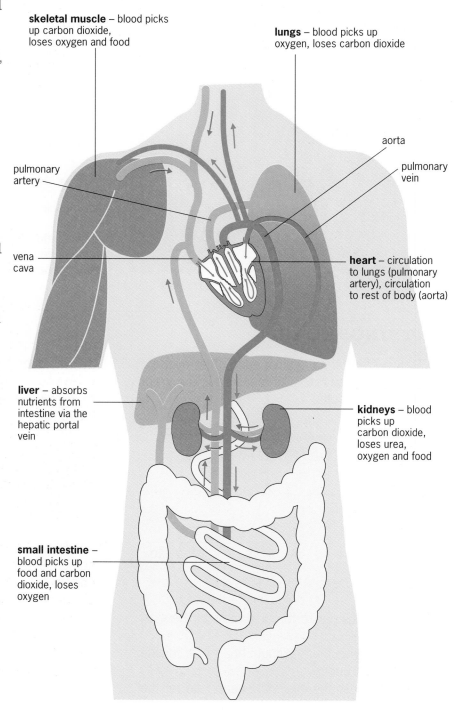

How the circulatory system works

HUMANS AS ORGANISMS

Blood vessels

Blood vessels are flexible tubes through which the blood flows around the body. Laid end to end, they would be over 1000 miles long. Some are arteries, others veins, yet others are called capillaries. When tissues are damaged, blood sometimes leaks out of these vessels. This leaked-out blood slowly changes colour from red to purple, then to yellow – forming a bruise.

1 The arteries carry blood *from* the heart to every other organ in your body. The blood is pushed from the heart under pressure and can make the artery walls stretch. The walls are thick and strong, to prevent them from stretching too much. They contain muscle and elastic fibres, and relax and widen as blood enters them, then spring back, helping to squirt the blood along

2 Arteries branch into smaller tubes called capillaries. Capillaries are only the thickness of a single cell and thread their way through every tissue and organ. They are so narrow that red blood cells can only just squeeze through. Their walls are 'leaky', so the liquid part of blood passes through the capillary walls, and flows around the cells, carrying substances that cells can absorb. Other substances pass back from the cells into the blood solution and then into the capillaries. The capillaries eventually join up into larger tubes called veins

3 Veins carry blood from the organs *back to* the heart. They have thinner walls and are wider than arteries, so offer less resistance to blood, making it easier to flow. But the pumping effect of the heart is lost, as the veins' thin walls lack muscle. Blood is slowly forced along veins as body muscles elsewhere contract, while valves inside the veins help keep the blood moving back towards the heart

Sections through an artery, a capillary, and a vein

★ THINGS TO DO

1 Look at the diagram opposite, which shows the main routes taken by blood as it flows around the body.
   Describe the routes by which the following pass through the circulation system:
   a) oxygen in the blood from the heart to the intestines,
   b) food in the blood from the intestines to the liver,
   c) water in the blood from the heart to the kidneys,
   d) oxygen from the lungs to where it is used in the leg muscle,
   e) carbon dioxide in the arm muscle tissues to the lungs.
   In each case, make a list of the blood vessels that the substance travels through, or draw arrows on a copy of the circulation layout diagram.

2 Design and make a 'circulation system board game'. The board could include the route through the heart, major organs and blood vessels. You could have rewards and penalties marked on the board or on card. What would be the aim of your game? How would you decide the winner?

3 Think about how substances are carried around your body in the blood, and how substances are carried around a city on the road system.

   a) How are the two systems similar?
   b) How are they different?
   c) Road systems sometimes have diversions. So, too, can parts of the body circulation.
   i) Why would it be helpful to divert blood to the leg muscles instead of the gut when you run quickly?
   ii) Why is it vital *not* to divert blood from the brain?

1.14 Life blood

Desperate appeal for blood as faulty bags are discovered

Hospitals throughout the country were today checking their stocks of blood after one hospital discovered that some bags that they were using had faulty seals. At low temperatures blood can be stored for up to 5 weeks. Most blood is used to replace that which patients lose during routine operations. Other stocks may be rapidly used during emergencies.

The bag with a faulty seal was discovered by a keen-eyed nurse as it was being transferred from the blood bank to the operating theatre. A specialist confirmed that microbes could enter blood through the faulty seal, and be transferred to the patient during transfusion. Patients treated with blood that has been exposed to air could end up with severe blood poisoning.

Blood transfusion

During lengthy operations, patients may lose up to 2 litres of blood. This must be replaced so that it can supply the cells in the patient's body with the oxygen and food that they need. When someone is given a transfusion, a bag of blood from a blood bank is connected to a vein in the patient's arm. As the blood passes into the patient it mixes with their own blood. The right type of blood must be used; people need blood that matches their own. The wrong type could lead to fatal complications.

The table shows which blood groups can be mixed in transfusions.

Blood group of receiver	Donor blood group			
	A	B	AB	O
A	✓	✗	✗	✓
B	✗	✓	✗	✓
AB	✓	✓	✓	✓
O	✗	✗	✗	✓

✓ = Safe   ✗ = Dangerous; mixed blood goes lumpy.

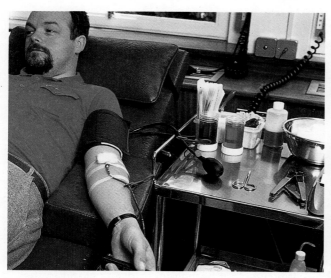

In a normal week from 40 to 120 units of blood are needed by the average hospital (1 unit = 350 cm^3). This blood is supplied by donors, who give about half a litre of their own blood each time they visit a blood donor session. The blood is then checked for any disease that might be present. If it is free from disease it is stored at low temperatures in a blood bank until needed

Blood will have to be transfused at the scene of an accident if an emergency operation has to be carried out to save someone's life or stabilise them for the journey to the hospital

HUMANS AS ORGANISMS

What is blood?

What is in a blood transfusion bag

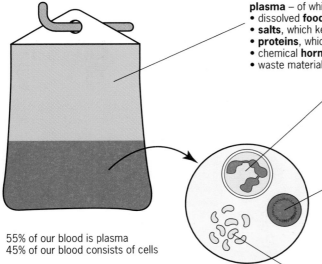

plasma – of which 90% is water, and the other 10% includes:
- dissolved **food**, such as sugar and amino acids,
- **salts**, which keep the blood at the right concentration and help the cells to work properly,
- **proteins**, which help the blood to clot and also help protect the body against invaders,
- chemical **hormones**, which control the way the body responds,
- waste materials, such as carbon dioxide and **urea**

white blood cells – approx. 7000 mm^{-3}; work together to protect the body against disease. They can kill microbes or foreign cells by engulfing these and digesting them. When we are unwell their number increases significantly. The different types can be identified by the shape of their nucleus. Some are made in the lymph glands, and some in yellow bone marrow

red blood cells – approx. 5 million mm^{-3}, these must be made continuously in the red bone marrow because after 4 months the liver destroys them. They have no nucleus, allowing more space for a substance called **haemoglobin**. Oxygen 'sticks' to haemoglobin as red blood cells pass through the lungs (forming oxyhaemoglobin). Then when the blood passes through the organs of the body the oxyhaemoglobin splits up, releasing its oxygen, which can then be absorbed by the cells that need it

platelets – approx. 0.25 million mm^{-3}; these are made in red bone marrow and are actually fragments of cells, but with no nucleus. They are needed to make a blood clot (e.g. if you cut your finger)

55% of our blood is plasma
45% of our blood consists of cells

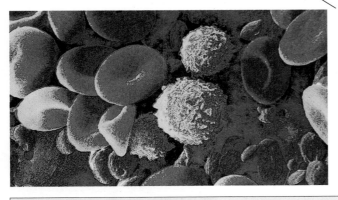

False-colour scanning electron micrograph of human blood cells (red blood cells = red; white blood cells = white; platelets = blue) (magnification ×25 300)

★ THINGS TO DO

1 Andy Smith was admitted to the casualty department of a hospital after a road accident. He needed blood urgently.

A test showed that Andy's blood group was group A. His father and mother immediately offered to donate blood so that Andy could be given a transfusion. Use the information in the table to decide which of these people's blood could be used for Andy.

- Andy's father: blood group B
- Andy's mother: blood group AB
- An anonymous donor: blood group O

2 Use the photographs of blood cells above to answer these questions.

a) How is a red blood cell different from a white blood cell?
b) How many red blood cells are there compared with white blood cells?
c) What jobs do they do?
d) How do the differences suit them to their different jobs?
e) How big is a white blood cell?
f) How big is a red blood cell?
g) Anaemia is a condition caused by having too few red cells in the blood. Why would you feel tired out if you had anaemia?
h) What problems would your body have if your blood contained fewer than normal white cells?

1.15 Invaders

How are diseases spread?

The tiny organisms that cause disease – called **microorganisms** (microbes) – live in, on and around us. They are what we call 'germs'. Many are harmless, and some are useful, but others cause diseases that interfere with the way the body works. That's when you feel ill.

Even if you come into contact with germs you may not catch the disease. You will only be affected if they manage to enter inside your body. The diagram shows how this can happen.

breathing in air infected with microbes (possibly by a diseased person sneezing)

drinking infected water (this passes on diseases like cholera and typhoid)

eating contaminated food (can lead to food poisoning)

entry through natural openings (such as the mouth, ears, eyes, rectum, penis or vagina)

entry through a cut, graze or sore (which allows them to invade the tissues beneath the surface, or the blood)

touching materials already used by an infected person (e.g. using a towel after a person who has chicken pox)

How germs get into the body

Different germs

Bacteria, **viruses**, single-celled creatures called **protozoa** and fungi (see Topic 4.1) are different types of microorganisms that cause disease. They are most likely to do this if large numbers enter the body from an infected person or because of unhygienic conditions.

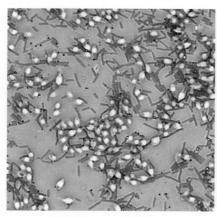

Each bacterium is a single living cell. The cytoplasm and membrane are surrounded by a **cell wall**. Unlike human cells, the **genes** (see Topic 4.4) are not contained inside a clear nucleus. Bacteria can reproduce very rapidly in the warm conditions provided by the body

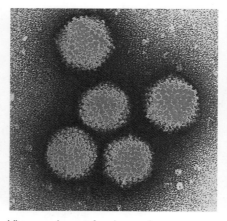

Viruses do not feed, respire or move, so they are not like other living organisms. They are smaller than bacteria and reproduce only inside the living cells of other living things. The few genes in the virus are surrounded by a coating of protein

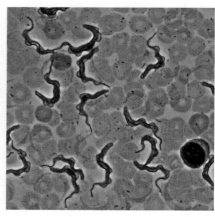

Protozoa are single-celled organisms with a nucleus – these include trypanosomes, which cause 'sleeping sickness' in tropical countries; here trypanosomes are seen infecting blood cells

HUMANS AS ORGANISMS

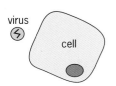
1 The virus approaches the cell membrane

2 The virus fastens itself to the cell and a 'string' of genes is transferred to the cell interior

3 The complete string of genes multiplies rapidly

4 A new virus is formed around each string of genes

5 Eventually the cell membrane bursts open and the new viruses escape

How a virus invades a cell and multiplies

The diagram shows what happens when a virus gets inside the cells inside your body. In this way viruses destroy the cells in which they reproduce. Some also produce poisonous chemicals – **toxins** – which interfere with the way the body works. Most viruses cannot be destroyed by drugs but only, in time, by the white blood cells.

Sealing a wound

Losing blood through a wound can be fatal because:
- vital substances like oxygen don't reach the organs effectively,
- microbes enter blood and are carried around the body.

Fortunately, blood quickly solidifies (clots) when exposed to air.

Clotting is caused by a series of changes that are speeded up by enzymes and helped by blood platelets, causing a soluble protein in the blood to be changed into insoluble strands of protein fibre. The fibres bind together to form a web, which traps red blood cells. These form a solid lump, which quickly seals the wound. After a few days, new cells grow over the wound to make the repair permanent.

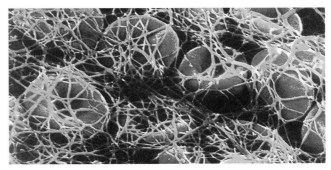
Fibres and red blood cells in a clot

★ THINGS TO DO

1. Explain how each feature below can reduce the risk of germs entering your body:
   a) sneezing into a handkerchief, not into the air,
   b) adding chlorine to water in a swimming pool,
   b) keeping food cool in the fridge,
   d) cleaning wounds with disinfectant and covering with a plaster,
   e) using an electric hand-drier instead of a towel after washing your hands in a toilet.

2. The faeces of cats may contain a protozoon called *Toxoplasma gondii*. This can cause a disease called toxoplasmosis, which leads to eye infections and blindness in people accidentally touching contaminated dirt.
   a) Prepare a leaflet telling cat owners of the problem and think of helpful advice on reducing the risk. What precautions could be used to avoid passing on the organism?
   b) Which groups of people should be especially careful?

3. Some people (haemophiliacs) have blood that does not clot properly. Their blood lacks an enzyme needed to form insoluble protein fibres.
   a) Explain how this could be life threatening.
   b) How would they be helped by supplying the missing enzyme in a blood transfusion?

1.16 Fighting disease

The diagram shows how your body uses two sets of defence mechanisms to:

- prevent harmful microbes (germs) entering,
- kill any germs that do get into the blood

Hungry for germs

If your body is invaded by bacteria, such as when you get a bad cut or a sore throat, you make more white blood cells to fight the invaders. Lymph glands in your armpit may swell up as extra white blood cells are produced.

Extra support

Microbes can often multiply quicker than white blood cells can destroy them. Chemicals called **antibodies** are needed to fight these microbes. Antibodies are made by the immune system (white blood cells and lymph glands). Their production and action is called an **immune response**.

a The first line of defence; **b** the second line of defence

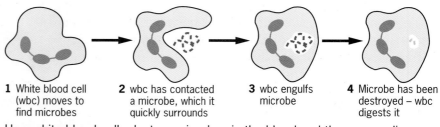

How white blood cells destroy microbes in the blood and the surrounding tissues

How the immune response works

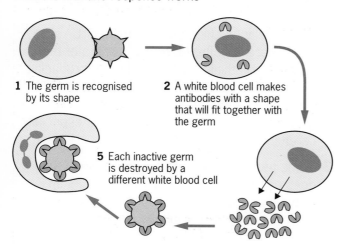

Different types of antibody are produced to fight different types of germ. Also, cells can remember what types of antibody they have already made. So when a particular sort of germ re-enters the blood it is immediately destroyed by newly made antibodies, which are produced very rapidly this time. This 'learnt' immune response tends to prevent diseases like chicken pox happening twice.

A history lesson

At the end of the eighteenth century almost everyone in England caught a disease called smallpox. Over one-third of the population died from its effects.

HUMANS AS ORGANISMS

But in Asia few people caught smallpox. The Chinese had known for thousands of years how to protect people by deliberately injecting (**inoculating**) them with a weakened form of the virus which causes smallpox.

In 1771 an English doctor called Edward Jenner noticed that people who worked with cows didn't catch smallpox, though they did catch a milder disease called cowpox. In 1796 he deliberately infected a young boy called James Phipps with cowpox by scratching some liquid from a cowpox blister on to the boy's arm. A few weeks later he tried infecting the boy with smallpox from a patient's blister. The boy did not catch smallpox.

Edward Jenner

This idea of protecting people from smallpox quickly caught on, and by 1840 vaccination was compulsory by law in Britain. Since then a **vaccine** containing a harmless form of the smallpox virus has been used to make smallpox extinct.

Vaccines against other germs are now given to children to reduce the risk of disease. This is called **immunisation**.

Age of child	Disease immunised against
3–6 months	diphtheria, whooping cough, tetanus, polio
4–6 weeks later	second diphtheria, whooping cough, tetanus, polio
4–6 weeks later	third diphtheria, whooping cough, tetanus, polio
1–2 years	measles
4–5 years	tetanus, polio, measles
10–13 years	tuberculosis
11–14 years	rubella (girls only)
15 years	tetanus, polio

Repeating the dose of vaccine ensures that the immune system remembers what type of antibody to make. So the immunisation effect is longer lasting.

★ THINGS TO DO

1 a) What idea was Edward Jenner investigating?
b) What conclusion did he come to?
c) Why did he test his idea on a healthy 8-year-old boy?
d) Why would he be banned from doing his experiment today?

2 This graph shows how immunisation has led to a reduction in the number of child deaths occurring from whooping cough.
a) Write down as much as you can about what the graphs show.
b) It could be argued that the information on the graphs was not reliable enough to be able to say that the deaths from whooping cough decreased because the number of children immunised increased. What other explanations could there be for the number of deaths decreasing?

c) What other information would help you make a more informed decision?
d) Try to collect information on immunisation against other diseases and make a display. You might find facts from a health centre, or even a travel shop!

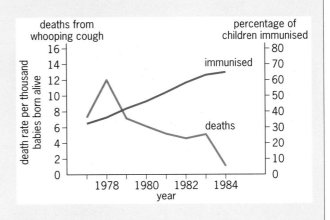

1.17 Lose it

The body makes many different chemicals, some of which are harmful waste materials. Waste like carbon dioxide, urea (made by the liver from surplus protein) and even excess water must be removed before they damage your body. Getting rid of harmful waste is called **excretion**. The organs that do this job make up the **excretory system**.

Food that has not been absorbed from the gut is also removed from the body. This process is called **egestion** (not excretion), because the waste substance has not been made by the body.

Problems can quickly emerge if the excretory system stops working effectively. This is especially the case if the kidneys go wrong – kidney malfunction can kill.

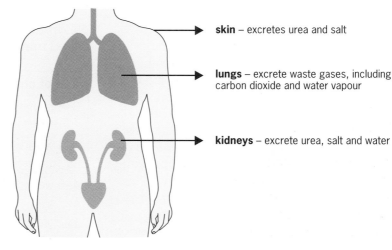

skin – excretes urea and salt

lungs – excrete waste gases, including carbon dioxide and water vapour

kidneys – excrete urea, salt and water

The excretory organs

Superdeal is off!

Athletico Italio have just called off their transfer of Liam O'Casey from Wynport United. United boss Bill Shankham explains: 'The lad is worth a mint, but he failed a medical. Apparently he's only got one kidney. He must be sick.'

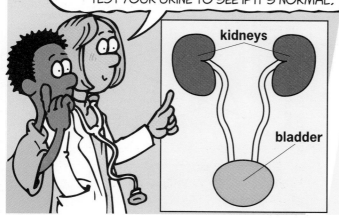

HUMANS AS ORGANISMS

How the kidney works

Each kidney receives blood from a **renal artery**; this splits into tiny capillaries inside the kidney. The liquid part of blood, containing water, urea, sugar and salt, filters out of the capillaries into microscopic tubules (**nephrons**) in the cortex. The useful parts are then reabsorbed (see Topic 1.18). What's left inside the tubule (water, urea and a little salt) is called **urine**. It is made all the time, and passes to the bladder to be stored; urinating gets rid of the harmful urea and any water that your body does not need.

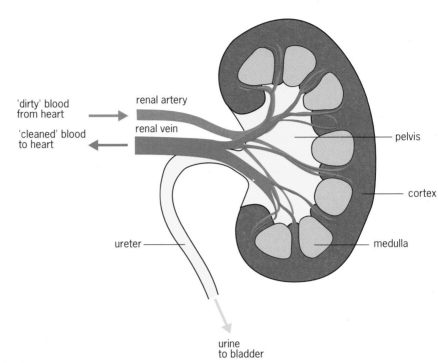

A section through a human kidney

Analysing urine

The kidneys can change the amounts of different substances in urine so that wastes are removed effectively. When you drink a lot, more urine is made, so more water is lost. This keeps the amount of water in your body balanced. The same happens with salts. The concentration of salt ions in urine depends on how much salt you have eaten and how much you have lost in sweat.

Urine can be tested to see how well the kidneys and other body organs are working. Chemical tests and looking at urine under a microscope may reveal particular problems.

For instance, people with diabetes insipidus make lots of urine, but people with high fever pass little. The urine of leukaemia patients contains a lot of uric acid, whereas sugar in it is a symptom of diabetes mellitus. Hepatitis, a liver disease, causes dark urine (from the presence of bile salts). Protein in urine suggests kidney damage, while red blood cells in it may indicate that a cancer is present. Pregnancy can also be detected by testing urine.

★ THINGS TO DO

1 In Victorian times a doctor would check a sample of urine for sugar by tasting it!
  a) What medical condition would the doctor be looking for?
  b) Ask your teacher if you can do a chemical test for sugar in 'urine samples' (not *real* urine).

2 If possible, examine a fresh kidney and identify the main parts. Draw a diagram and label it.

3 Explain how urine tests can be used to detect:
  a) drunken drivers;
  b) athletes taking banned drugs;
  c) pregnancy.

1.18 A filter for life

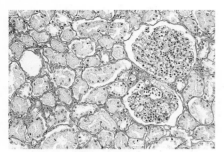

(above) Each kidney contains about 1.25 million nephrons, which are filtration units where urine is made; (right) how a nephron works

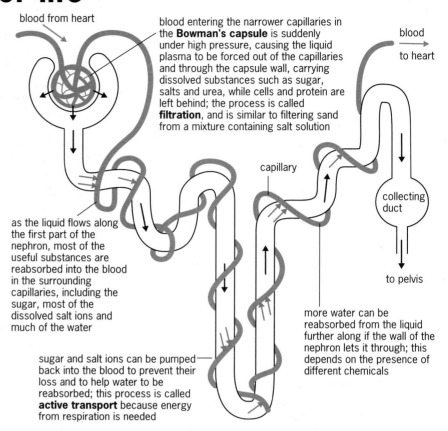

blood from heart

blood entering the narrower capillaries in the **Bowman's capsule** is suddenly under high pressure, causing the liquid plasma to be forced out of the capillaries and through the capsule wall, carrying dissolved substances such as sugar, salts and urea, while cells and protein are left behind; the process is called **filtration**, and is similar to filtering sand from a mixture containing salt solution

blood to heart

capillary

collecting duct

to pelvis

as the liquid flows along the first part of the nephron, most of the useful substances are reabsorbed into the blood in the surrounding capillaries, including the sugar, most of the dissolved salt ions and much of the water

sugar and salt ions can be pumped back into the blood to prevent their loss and to help water to be reabsorbed; this process is called **active transport** because energy from respiration is needed

more water can be reabsorbed from the liquid further along if the wall of the nephron lets it through; this depends on the presence of different chemicals

One-fifth of all the blood pumped by the heart per minute in a healthy person goes through the kidneys. By altering the conditions around each nephron the composition of urine can quickly be changed. A hormone called **ADH (anti-diuretic hormone)** makes the nephron wall leaky, so water passes back into the blood very easily. Less water remains in the urine, so it is more concentrated. The effect is called **anti-diuresis**.

Without ADH, water stays in the urine, and more, dilute urine is made. The effect is called **diuresis**. Chemicals such as alcohol have this effect. Doctors use drugs called diuretics to help patients with kidney failure or low blood pressure to make more urine.

Kidney problems

A kidney stone can stop urine flowing out of a kidney and cause pain. It can be broken up by ultrasound treatment. A kidney infection is detected by looking for microbes in urine; **nephritis** (kidney inflammation) is the commonest form of kidney disease. It can lead to kidney failure, in which the kidneys stop working. One response to this life-threatening condition is to put the patient on to a kidney dialysis machine. A longer-term solution is a kidney transplant.

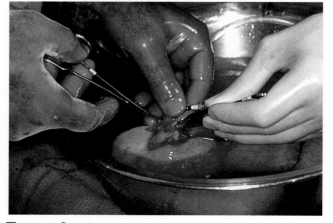

Transplant
Transplanting a kidney is usually a successful operation. But difficulties arise if the patient's immune system recognises the new kidney as a foreign object and tries to destroy it. To reduce the risk of this happening, the blood of the new (donor) kidney is matched as closely as possible to that of the patient (recipient); genetically engineered organs may also be used (see Topic 4.17), and immunosuppressive drugs to reduce the activity of the patient's immune system.

HUMANS AS ORGANISMS

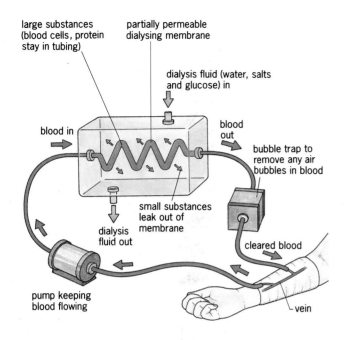

Dialysis

During dialysis, blood passes from a patient's artery into a machine called a dialyser (artificial kidney). This keeps blood at body temperature as it passes between a series of plastic tubes that are partially permeable (i.e. some substances can leak through) and contain dialysing fluid.

This fluid has a similar composition to blood, so that only waste substances and *excess* salts, sugar and water pass from blood into the dialysing fluid. The 'cleaned' blood then returns to the body through a vein. A patient may be on a machine for up to 8 hours, three times a week.

Dialysis can be used to clean a patient's blood unitl the kidneys recover, naturally or as a result of surgery or treatment. But ultimately most patients would benefit from receiving a transplanted kidney.

★ THINGS TO DO

1 Look at the information on dialysis.
  a) Why do you think the dialysing fluid is kept at body temperature?
  b) Why must germs be kept out of the machine?
  c) What substances must not be lost from blood during dialysis?
  d) Kidney failure can be treated by dialysis or transplant. List the advantages and disadvantages of each treatment.

2 Kidney damage can be analysed by injecting a radioactive substance into a patient's blood and measuring the amount of radiation given off by each kidney. A normal kidney gives off more and more radiation for up to 10 minutes as the radioactive substance is filtered into the nephrons. The level then drops to zero as the substance is passed out of the kidney into urine.

The results for one patient are shown (right).
  a) Plot the data as two lines on to a graph, showing the amount of radiation against time.
  b) Which is the normal kidney? Explain why you think this.
  c) What seems to be happening in the other kidney?

Time/ min	Amount of radiation	
	Left kidney	Right kidney
1	16	10
2	23	25
3	33	39
4	41	38
5	49	20
6	56	55
7	65	19
8	79	59
9	70	23
10	60	55
11	54	8
12	42	44
13	33	35
14	24	77
15	15	15

1.19 Feeling sweaty

John is training to be a jockey. Daniel has been a jockey for 10 years and is explaining why he often has a sauna.

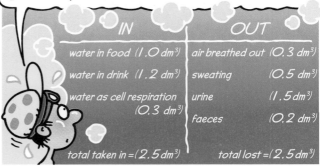

Skin

The skin is an organ which acts as a barrier, helping or preventing substances entering or leaving the body. It helps to balance both the amount of water and the body temperature.

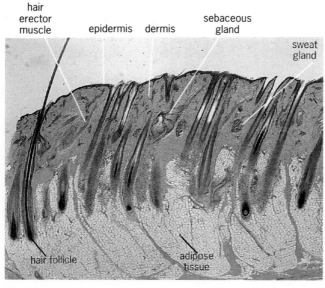

What hairy skin looks like under a microscope

When you are hot, sweat glands in the skin make **sweat**, a solution of salt and urea in water. The water in sweat evaporates, leaving the skin cooler. This in turn cools the blood in capillaries in the skin. The cooled blood is pumped around the body, cooling any overheating organs inside. The skin also loses heat by radiation and by conduction.

To increase the cooling effect, more blood is pumped closer to the skin surface. This effect is called **vasodilation**. When the body temperature returns to normal (37 °C) you stop sweating and less blood is carried to the skin; this is called **vasoconstriction.**

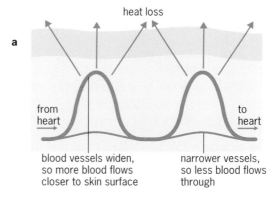

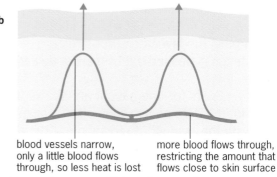

Blood vessels: **a** vasodilated; **b** normal

The more you sweat, the more you cool down; but you also lose more water, salt and urea. Urea is a waste material, so it does not need to be replaced. But if you lose too much water and salt you will suffer from dehydration. The amount of water and salt lost needs to be replaced by drinking and eating.

In direct sunshine this person may lose as much as 500 cm³ of sweat each hour

In the shade less sweat is produced, but still as much as 250 cm³ per hour may be lost

★ THINGS TO DO

1 a) Copy the information Daniel wrote in the sauna into your book. Use it to draw one pie chart showing water taken in and another showing water lost.
b) Explain how sweating helps stop Daniel becoming overweight.
c) The amount of water Daniel loses by breathing out will stay the same. Why?
d) Daniel will produce less urine after his sauna. Can you explain why?

2 Look at a skin section under a microscope or on a CD-ROM. How many sweat glands can you see? How does this compare with skin from different parts of the body?

3 Two pupils decided to find out what things affect how quickly a body cools down. They used conical flasks covered with clothing material.
a) What do you think will affect how quickly the body cools down? Make a list of your ideas, giving a reason for each.
b) Choose two ideas and plan how you could test them, including steps you will take to make your tests safe. When your teacher has checked your plans, carry out your investigations and then write them up in a report.

4
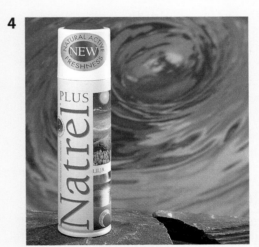
a) Anti-perspirants are used to 'keep you dry'. How do you think they work? How could you test this idea? Plan your test, have it checked, then try it.
b) Prepare a short article for a magazine called 'How things work', explaining how anti-perspirants work. (You can use your test results to support what you say in the article.)

5 a) During a marathon race (40 kilometres) runners drink water and may take salt. Why is this important?
b) How would running on a very hot day affect them?
c) Why don't runners in a 1-kilometre race drink water as they race?

1.20 Staying warm

Hypothermia

Many people who take part in outdoor pursuits risk suffering from **hypothermia** – when the core body temperature falls well below 37 °C. This is why when people go out hill walking, or climbing, they are advised to wear sufficient warm clothing and to take provisions and equipment, so in an emergency they have shelter and food until help arrives. Elderly people and young babies are also at risk during cold spells during winter. The body temperature does not have to drop by much before you begin to suffer ill-effects. These effects are the result of enzymes in your cells not working effectively outside their normal temperature range.

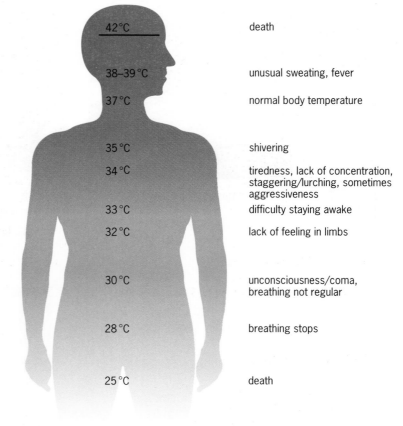

Effects of changes in body temperature

Warm-blooded animals

Humans, like other mammals and also birds, can keep the inside of their bodies constantly warm. The normal 'core' body temperature for humans is 37 °C (but the skin can be slightly cooler).

The body responds to cold conditions by:

- reducing the amount of blood flowing into the skin, so less heat is lost by radiation,
- the hairs on the skin sticking up to trap an insulating layer of air,
- closing the sweat pores to reduce loss,
- producing more heat to warm the blood by extra respiration in the liver,
- altered behaviour, in ways that keep you warm (e.g. putting on extra clothing, more movement).

In very hot conditions the opposite responses are needed, so:
- more blood flows into the skin,
- hairs are flattened to reduce insulation,
- sweat is released from the skin,
- the liver does not release extra heat by respiration,
- behaviour changes (e.g. less vigorous movement, wearing less clothing).

Excessive overheating, such as a high fever brought on by illness, can be dangerous. A rise of only 5 °C to 42 °C could be fatal. Some medicines, including aspirin, lower body temperature during illness. Other chemical substances (such as some illegal drugs) are dangerous because they allow the body to overheat.

HUMANS AS ORGANISMS

Some mammals **hibernate** in winter; this is like a long deep sleep in which the temperature is maintained at a much lower level than usual (down to 2 °C)

Animals such as snakes are cold blooded. Their body temperature is cooler than that of mammals and is affected by changes in their surroundings. Snakes often bask in the sun; this raises their temperature. During the cooler nights, they find shelter and coil up to retain as much body heat as possible

★ THINGS TO DO

1 Climbers, fell walkers and potholers are all in danger from hypothermia. Low temperatures, rain and wind all have the effect of speeding up heat loss from the body.
   a) Carry out an investigation to find out which of the above conditions has the greatest cooling effect on the body.
   b) If possible, extend your investigation by using computer-linked data-logging equipment.
   c) You could also test the relative effects of different materials on reducing heat loss.
   d) Write a clear account of your investigation, using the help given to you by your teacher.

2 **a)** Prepare a leaflet for elderly people giving them advice on how to avoid hypothermia. It should describe the symptoms of hypothermia so that they can recognise when they are at risk.
   b) Find out what advice is given to mothers on how to keep their newborn babies warm.

3

These penguins are keeping warm in the Antarctic. Their surroundings can drop to −40 °C, yet they still manage to keep their core body temperature at about 40 °C.
   a) Why should huddling together in a group help penguins to keep warm?
   b) How could you investigate your ideas?
   c) Test your ideas and write a report of what you find.
   d) Think of other ways to keep warm and investigate how effective they are.

1.21 Keeping a balance

> Secretary,
> Osteoporosis Society
>
> 14 Forest Side
> Longfield
> Essex
>
> I am writing to you because my grandma has just been told she has a condition called osteoporosis. The doctor says that her bones have become brittle. Can you explain why and what could be done to help her?
>
> Yours faithfully
>
> Claire

Osteoporosis is a particular problem for older women

> Dear Claire,
>
> Many thanks for your letter. I am sorry to hear your grandma has osteoporosis. This condition is increasingly common in older women. The problem starts when the body cannot control its calcium levels. The bones start to lose calcium, which helps to keep them strong. Calcium ions are absorbed into the blood, and not replaced from the diet. The bones get weaker and break more easily. It can be very painful, and the sufferer cannot move easily.
>
> Chemical messengers, called hormones, control what the bone cells must do to keep bones in good condition. A hormone called oestrogen usually makes sure that there is enough calcium in the bones. Without it, bones can become very brittle. Oestrogen hormone replacement therapy (HRT) can be used to treat the condition. Injections of it can help to strengthen the bones if foods or supplements containing calcium, magnesium and traces of boron, as well as vitamin D, are taken in the diet.
>
> A message to everyone is to drink plenty of milk and eat other foods that contain calcium, such as canned fish, or leafy greens. We often hear that it is important for growing children, to ensure that bones and teeth grow strong. It is also important to replace calcium losses as people age. Regular exercise can help, too.
>
> Yours sincerely
> Pamela Smith
> Bone Specialist

Constant conditions

Although the conditions around you are changing, the millions of cells in your body need constant conditions if they are to work properly. Body cells are surrounded by a liquid called **extracellular (tissue) fluid**, which contains blood plasma. They work best if this liquid contains the right concentrations of water and dissolved substances, such as calcium ions, and stays at body temperature. Blood also needs to do the same. The process of keeping internal conditions constant is called **homeostasis**, as the diagram shows.

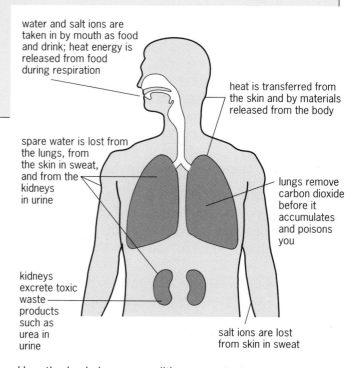

How the body keeps conditions constant

HUMANS AS ORGANISMS

In control

The processes of homeostasis are controlled by **hormones**. These are chemicals made in glands and released into the blood, which carries them around the body. Hormones act as messengers, telling organs what to do. But the message is only carried out by organs which can detect a circulating hormone. The message does not last for ever, either, as hormones are destroyed in passing through the liver.

Insulin and **glucagon** are two hormones released from the pancreas that control blood sugar levels (see Topic 1.22).

ADH is a hormone that tells the kidneys to reabsorb more water from urine (see Topic 1.18). The number of salt ions dissolved in the blood increases when salt is absorbed after a meal. A hormone called **aldosterone** tells the kidneys to remove salt, reducing the salt level to normal.

Another hormone controls the degree to which the kidneys reabsorb hydrogen ions back into the blood. An increase in the concentration of hydrogen ions makes the blood acidic. The body works best if the blood is slightly alkaline (pH 7.4).

Oestrogen is a sex hormone, but it also maintains the right amount of calcium in bones, to keep them strong. **Testosterone**, another sex hormone, has the same effect in males.

Hormones with different messages

Not all hormones are involved in homeostasis, but they are all circulated by blood around the body to carry messages.

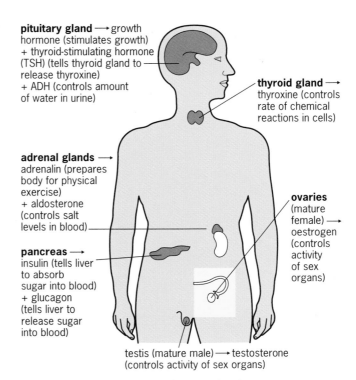

Hormone production by different glands

★ THINGS TO DO

1 The table shows the amount of calcium needed at different stages of life.

Age/years	Amount of calcium needed/mg
<12–15	1200–1500
15 onwards	800–1000
>45 (women)	1500
pregnant/nursing mothers	1200

a) Why is a lot of calcium needed in the first 15 years?
b) Why is less calcium needed after 15 years?
c) Children who lack calcium as they grow develop poor bones – a condition called rickets. Look back to Topic 1.5 to see what foods would provide the calcium they need.
d) What condition could older people develop if they lacked calcium?
e) Explain how a lack of oestrogen could also lead to this condition.
f) Why is homeostasis important?

1.22 Life in the balance

What is diabetes?

Diabetes mellitus, to give it its full name, is a condition that can be life threatening if not controlled. Both adults and children can suffer from it. People with diabetes (called diabetics) cannot effectively control the amount of sugar in their blood. The hormone insulin, which normally controls blood sugar levels and keeps them constant (see Topic 1.21), is not working properly.

The symptoms of diabetes are: thirst, frequent urination, tiredness and weight loss. Very high or low levels of sugar in the blood can lead to unconsciousness – a diabetic coma. As the level of sugar in the blood rises, the kidneys try to lose the excess sugar. So one of the signs of diabetes is sugar (glucose) in urine.

The condition cannot be cured, but most diabetics live normal, active lives without any of the symptoms described above. They control their blood sugar levels by:

- self-monitoring of their blood glucose levels,
- giving themselves regular injections of insulin, which makes the liver remove any excess glucose from the blood (this can now be made by genetic engineering – see Topic 4.17),
- avoiding eating too much sugar.

Up and down

Keeping the level of sugar in the blood steady is no easy job. The level rises after eating carbohydrate, and falls during exercise. The diagram shows how the two hormones, insulin and glucagon, work in opposite ways to balance the amount of sugar in your blood:

- insulin causes blood sugar levels to fall, as sugar is absorbed by the liver and converted into glycogen,
- glucagon causes blood sugar levels to rise as sugar is released by the liver after conversion back from glycogen.

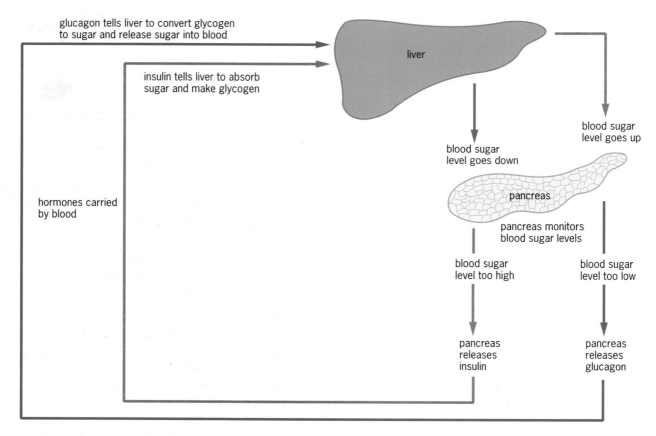

Keeping the blood sugar levels constant

Controlling the message

There are two mechanisms to stop insulin and glucagon passing their messages endlessly to the liver.

1. The liver slowly destroys the hormones, so each time blood flows through, less hormone emerges.
2. The hormones eventually become victims of their own success. For example, insulin causes a fall in blood sugar level. As the level returns to normal, the pancreas stops releasing insulin. This mechanism is called **feedback control**. Eventually, the insulin still in circulation is removed by the liver.

A job for adrenalin

Most hormones work slowly to maintain steady conditions. An exception is **adrenalin**, which is produced in the **adrenal gland**. In a dangerous situation you need to react quickly. Your whole body must respond by releasing the extra energy needed for a quick response: the liver must release extra glucose, the heart must beat faster, the lungs must work faster and the muscles must work quickly. Adrenalin stimulates all of

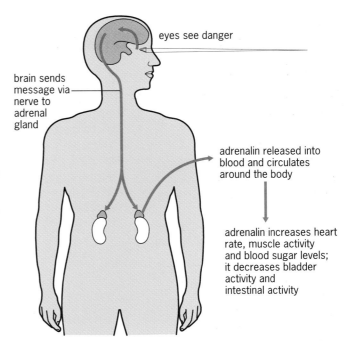

How adrenalin works

these organs into rapid action. It is also produced in worrying situations that are not physically dangerous (such as exams or social meetings).

★ THINGS TO DO

1.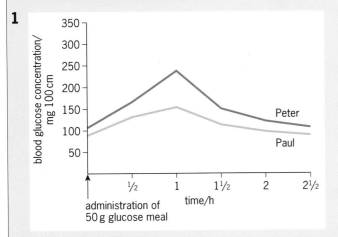

Peter was suspected of having diabetes. He had a special test, in which he was given a sugar meal then his blood glucose level was measured every half hour and his urine was tested every hour. (If the blood glucose level goes above 180 mg per 100 cm³, then glucose appears in the urine.)
Paul, who did not have diabetes, was also tested. The results are shown in the graph.
a) Why does Paul's blood glucose level rise for an hour after the meal?
b) Why did the blood glucose level drop afterwards?
c) Explain how the data on blood glucose levels show that Peter has diabetes.
d) When would Peter be making urine that contains glucose?

2. a) Explain what is meant by 'feedback control'.
b) Why is it important that glucagon is released only when blood sugar levels fall below normal?
c) Explain how feedback control works for glucagon.

1.23 Fast track

A racing driver needs a lot of skill. He must quickly sense and respond to changes. The speed and accuracy of his response will determine whether or not he survives the next corner

Dealing with changes

Conditions around us can change rapidly. To survive we need to sense change and to respond accordingly. The job is done by the **nervous system**. This comprises the **sense organs** (such as eyes and ears), the **brain**, **spinal cord** and **nerves**, which work together to sense changes, decide what these mean and tell different parts of the body called **effectors** (e.g. muscles, hormones) how to react – by moving parts or all of our body, for instance.

In the pits

Sometimes, as shown in the diagram, we need to react very quickly to avoid harm. In such cases, to reduce the time between sensing danger and reacting, the body cuts out part of the message system – the brain. This reduces the time for messages to pass from **sensor** to muscle, and so to move out of danger.

The response shown takes a fraction of a second, so any burn on the skin is minimised.

The nervous system, and a section across a nerve bundle

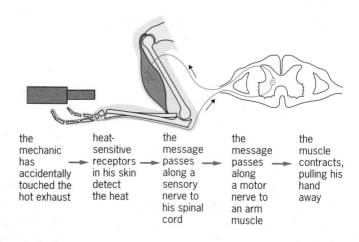

How a mechanic in the pits would respond to touching the hot car exhaust

HUMANS AS ORGANISMS

Nerve messages passing along a pathway that does not involve the brain are called a **reflex arc**, from sensor (skin) to effector (muscle). This can be summarised as:

sensor → spinal cord → effector

Messages involving the brain

At the same time as the above response happens, a different receptor in the skin may detect pain. This message, however, passes to the brain, and to then to the voice box and mouth, causing the mechanic to yell out! This pathway is not a reflex arc, and it is longer:

sensor → spinal cord → brain → spinal cord → effector

The message has further to go, so takes more time. So the mechanic yells out after he has removed his hand from the danger.

Most messages pass from the sensor to effector by way of the brain. Although it takes longer to do this, it is only by fractions of a second. The brain can then work out what the best response should be. A racing driver reaching a corner at high speed may either break hard or accelerate to drive around safely. The response to seeing the corner is not automatic, but thought out – it depends on the brain calling on experience and deciding what to do.

The brain can be trained, or conditioned, to respond to certain stimuli. You may not have realised how you react to everyday stimuli that have a particular meaning for you, for instance the sounds and smells of breakfast cooking, alarm clocks or school bells. Repeated practice or reaction reinforces particular actions controlled by the brain. This is used in animal training.

★ THINGS TO DO

1 There are many ways of measuring response time (how long it takes you to react to a change). The response in this case is not an automatic change because the brain is involved. You might find a way to measure response time on a computer. An alternative is to drop a ruler and catch it again, as in the diagram.

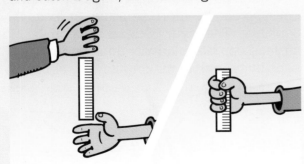

Work with a partner to investigate response time. You might find out if it:

- gets shorter with practice (over a few minutes or several days),
- is different for each hand,
- can be varied under different conditions, such as standing or sitting, in bright or dim light, with noisy distractions.

2 Copy the pattern of numbers 1 to 24 on to a piece of tracing paper. Look at the pattern using a mirror, as shown in the diagram. Try to join up the numbers in order, to make a picture, in the quickest possible time.

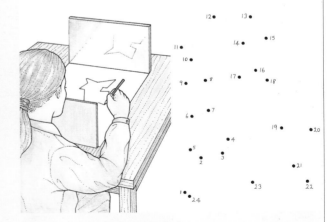

a) How long did it take?
b) Try again but this time look directly at the numbers (don't use the mirror). Explain any differences in times taken.
c) Can you learn how to improve your time?
d) Explain when it might be dangerous to take your senses and responses for granted.

47

1.24 Use your senses

Failure to react can be fatal! Being sensitive to changes around you helps to keep you alive. Different sense organs have a range of senses to monitor changes in the surroundings. Sensory cells in these organs detect specific changes, or stimuli, in the environment. Once a change is sensed, messages travel along the sensory nerves to the spinal cord, and usually the brain. A response is then coordinated.

Lives are lost in fires like this. You have only seconds to detect a fire and escape to safety. Seeing flames, smelling burning, hearing a smoke alarm and feeling heat are all important signs that you should get out quick

eyes – sense = vision (sight); receptors are sensitive to light

ears – sense 1 = hearing; receptors are sensitive to sound vibration
sense 2 = balance; receptors are sensitive to changes in body position

nose – sense = smell; receptors are sensitive to chemicals

tongue – sense = taste; receptors are sensitive to chemicals

skin – sense = touch; receptors are sensitive to changes in pressure and temperature

The body's sense organs

Mixing sensory cells

Taste is detected by a combination of sensory cells on the tongue. These cells are grouped into **taste buds**. Different taste buds detect different tastes – salt, sour, sweet, or bitter. Some foods are a blend of the four basic tastes. Others, like vinegar, have only a single taste. The brain works out the taste of a food from messages sent by different taste buds.

A sensory organ

The skin also contains a number of different sensitive cells. Skin in different parts of the body can have a different number of sensory cells, so can be more sensitive or less sensitive. The layer of dead cells on the skin surface can be relatively thick and reduce sensitivity or thin, making it more sensitive.

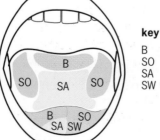

key
B = bitter
SO = sour
SA = salt
SW = sweet

Distribution of taste buds in the tongue

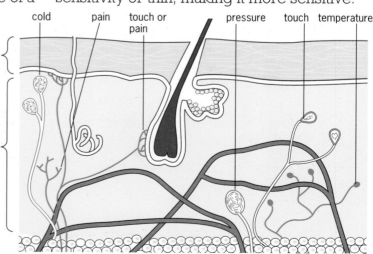

Sensory cells in the skin

HUMANS AS ORGANISMS

The ear

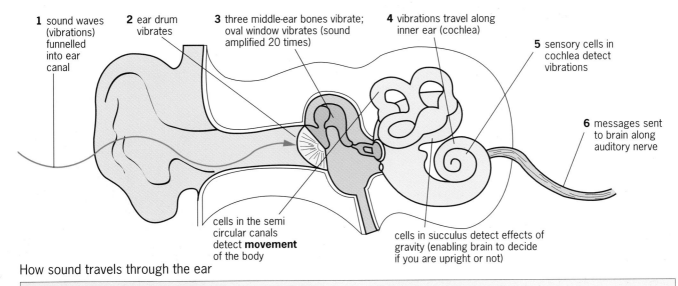

How sound travels through the ear

★ THINGS TO DO

1. You are probably familiar with the 'Take the cola' test! Tasting six different brands of cola to tell them apart isn't easy in the following conditions:

   - cola is cooled to 8 °C,
   - drink is served in a plastic cup,
   - cola is poured from the bottles into the cups at the same time,
   - lights are dimmed so any differences in colour are not obvious.

   a) Why would these conditions help to make the tasting test fair?
   b) What else might be important to make the test fair?
   c) How could health risks be removed from the testing method?
   d) Why might it be helpful for a number of people to take the test?
   e) If possible, do the test with a number of volunteers. Write a 'newspaper article' of your test and findings.

2. Some parts of your body are more sensitive to touch than others.
   a) Explain why the structure of skin could cause different sensitivities.
   b) Why might it be important for some parts to be more sensitive than others?

   c) Think of ways that you might test the sensitivity of different parts of the arm to touch. One idea would be to use a card pierced with two blunt pins 1 cm apart. By lightly touching on the arm of a friend, who is not looking, with one or both pins, you could discover how sensitive the skin was in different areas. The reliability of the results would be improved by touching the skin many times.
   i) Plan your test carefully, and have it checked, then do it.
   ii) Write a full report of your investigation.

3. **a)** Try a test with a radio or CD player to find out who in the group can hear the quietest sound. Think of what variables you should control and how you might do this.
   b) Write a report of your test.
   c) How could you test the hypothesis that hearing is poorer in older people because the sensitive cells in the ear die with age?
   d) Sound-sensitive cells in the ears can be destroyed by very loud noises, over 120 dB. If possible, use a sound meter to survey sound levels. You might investigate the level of sound at different distances from a road. You could test the effect of different materials in reducing sound levels.

1.25 A1 vision

The eye is a complicated organ. It is designed so that light rays are brought together (converged) and focused on to a layer of light-sensitive cells at the back of the eye (the **retina**). Light rays from an object are bent as they pass through the **cornea** and **lens**. This causes them to cross over, and the image of the object to be formed upside down (inverted). This image is detected by light-sensitive cells (**rods** and **cones**) in the retina, which pass messages to the brain along the **optic nerve** (see Topic 1.26).

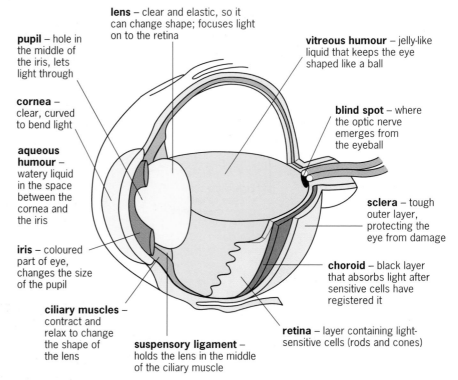

Inside the human eye (horizontal section)

Near or far?

As you read this book your eyes focus on it, a near object. But you will not at the same time be able to see things beyond the book clearly – they are blurred. If you then look up to see something clearly in the distance, the book will in turn be blurred.

To be correctly focused on the retina, light rays reaching the eye from a near object must be bent, or converged, more inside the eye than those coming from a distant object. So the shape of the lens must be able to change rapidly to accommodate light from different distances, as the diagram shows.

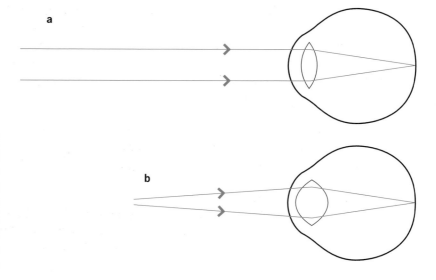

How the eye focuses: **a distant object** – the ciliary muscles contract (get shorter) and pull on the lens, so the lens is thinner and light is converged less; **b near object** – the ciliary muscles relax, so do not pull on the lens, which springs back into a more rounded shape, so light is converged more

HUMANS AS ORGANISMS

Light counts

The amount of light entering the eye depends on how big the **pupil** is. The size of this hole is controlled by the iris. Muscles in the **iris** contract and relax to make the pupil bigger or smaller. In bright light the pupil is made smaller. Too much light entering the eye could damage the retina and make the image less clear. The pupil enlarges when there is little light, so that more of the light-sensitive cells in the retina are stimulated.

Rods in the retina work where there is little light, so they let you see at night. Cones need more light, so work only in the day or artificial light. They detect different colours (see Topic 1.26).

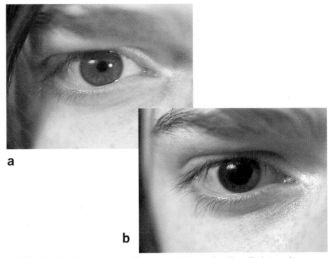

a Bright light reduces the pupil size; **b** dim light enlarges the pupil

★ THINGS TO DO

1

Photographs like this, taken with a flash, show 'red eye'. It is caused by reflection of some of the light from the white flash by blood vessels in the retina

a) Which part of the eye is red in the photograph?
b) What colour is this part in daylight?
c) Explain why it becomes red in flash light.
d) Some cameras try to reduce 'red eye' by using a small flash just before the main flash. The small flash makes the pupil smaller. How will this help?

2 **a)** Design and try out an 'eye test' with a few friends. For instance, you could read letters on a card or numbers on a car registration plate. Make sure that you control variables such as distance to make the test fair. Find out who has the best eyesight. How do the left and right eyes compare?
b) Why is an eyesight test an important part of a driving test?

3 If you look at the eye diagram opposite you will see a region called the **blind spot** on the back of the eye. This region has only nerve cells taking messages from the retina to the brain.
a) Why won't anything be seen in this region?
b) Follow the instructions below to find your own blind spot. Try to draw a diagram with light rays to show what happens when the dot disappears.

Hold the book up at about 30 cm away from your face. Close your left eye; now stare at the cross with your right eye. Without moving your eye, slowly bring the book nearer to your face. Before long the dot will disappear

1.26 Passing the message

Seeing isn't always believing. The image of the object you see forms upside down on the retina at the back of each eye. The brain has to make sense of that image by turning it the right way round. What you 'see' depends on what your brain is able to understand.

For instance, different receptor cells in the eye sense each of the 'primary' colours separately; it is the brain that combines their messages to arrive at the actual colours of the object seen.

The brain

The brain is a vital organ; we cannot live without it. It has three main jobs:

- to receive messages from sense organs about changes inside and outside the body,
- to coordinate responses by telling muscles or hormone glands (effector organs) what to do,
- to store and retrieve information so that future responses take account of previous experience (learned behaviour).

The brain is made of millions of nerve cells, with many connections between them. Blood provides a constant supply of the oxygen and food needed for its cells to be active and remove the wastes made. Nerve cells in the brain are connected to those in the rest of the body by the spinal cord. The skull protects the brain from physical damage.

Different parts of the brain have different jobs, as the diagram shows.

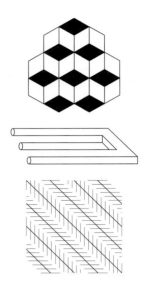

When you look at optical illusions, your eyes see what is there, but it may take some time for your brain to make sense of it

How we see colours

The eye has two main types of light-sensitive cells. **Rods** are sensitive only to white light. **Cones** are sensitive to different colours of light: some sense red, others sense blue and a third type senses green. The brain determines the colour of an object by combining the messages it receives from the cones. The diagram shows how mixing the three primary colours, red, blue and green, provides other colours, including white.

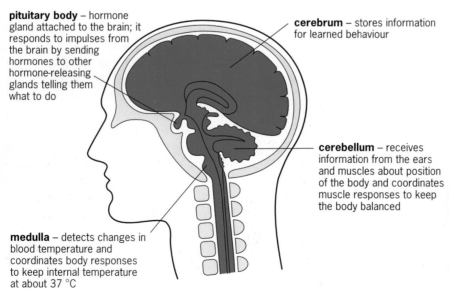

pituitary body – hormone gland attached to the brain; it responds to impulses from the brain by sending hormones to other hormone-releasing glands telling them what to do

cerebrum – stores information for learned behaviour

cerebellum – receives information from the ears and muscles about position of the body and coordinates muscle responses to keep the body balanced

medulla – detects changes in blood temperature and coordinates body responses to keep internal temperature at about 37 °C

Functions of different brain areas

How the three primary colours make other colours

HUMANS AS ORGANISMS

A nerve cell

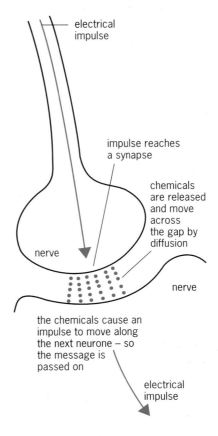

How a nerve impulse travels across a synapse

Nerve cells (neurones)

Neurones are specialised cells that allow messages to pass along them very quickly, to the end of their very long 'arms'. The message is in the form of an electrical impulse that travels along the cell membrane. A surrounding layer of fatty material (the **myelin sheath**) speeds the impulse up. Finger-like structures at the end of the cell make a close connection with other neurones. But neurones do not touch; a tiny gap called a **synapse** separates them. The synapse ensures that the impulse can travel only one way, but it also slows it down.

★ THINGS TO DO

1 Look back at the test to detect the blind spot in your eye (Topic 1.25).
   a) Copy the diagram on to several different-coloured papers (red, yellow, blue etc.) and then redo the test. Describe and explain what you see.
   b) Draw a vertical line through the spot. Redo the test.
   i) What do you see this time?
   ii) How do the eye and the brain combine to produce this effect?

2 a) What is a synapse?
   b) Why is a synapse important in a nerve pathway?
   c) Certain drugs can interfere with the diffusion of chemicals across synapses. Some stimulants make synapses work more easily. Why will this make the nervous system more active?
   d) Tranquillising drugs have opposite effects to stimulants. Explain how they might work.

3 Find out how 'magic eye' pictures work. Explain why some people are much better at seeing these images than others.

1.27 Drugs can harm

Dad sent to jail after steroid attack
Hampton magistrates today sentenced Ron Bin to 3 years' imprisonment for violence. This followed his repeated overuse of steroids in body-building sessions. The magistrate warned of the dangers of taking steroids. 'This type of abuse can lead to unpredictable bouts of uncontrolled aggressive behaviour. It has to stop.'

The word **drug** is used to describe any chemical substance that changes the way the body works. Medicines are useful drugs that improve health or help fight disease. Drugs like these kill microbes, readjust faulty processes in the body, remove pain and reduce a high temperature in illness.

But all drugs can have harmful side-effects. Drug companies are legally obliged to test drugs to see that they are safe and to discover side-effects. Doctors also look out for any harmful effects that a drug may be having on a patient, and change them to other types of treatment if necessary.

Some drugs affect the nervous system, particularly the brain, changing mood and behaviour. Caffeine (in tea, coffee and cola), nicotine (in tobacco) and alcohol (in beer, wines and spirits) are legal drugs that in small quantities can have beneficial effects. But overuse can bring health problems.

Athletes must be very careful about what they eat and drink. Even things like cough medicine and herbal teas may contain banned substances. Substances that are outlawed give the athletes unfair advantages and are a possible risk to health. Drug testing can trace chemicals in the blood, and cheats are caught and punished.

Like other drugs, steroids can have good uses. They may be taken by people suffering from asthma, or to help undersized children grow taller. However, they may be abused by people using them for non-medical purposes.

Many drugs have long been illegal because of their risk to health. These include 'hard' drugs, such as cocaine and heroin, and 'soft' drugs, like cannabis, Ecstasy and LSD. It is also illegal to sell or give to young people under 18 any solvents such as glue, lighter fuel and aerosols, which can cause intoxication.

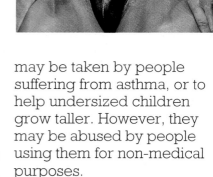

A recent poster from the Drugs Prevention Initiative

HUMANS AS ORGANISMS

Addiction

If a person takes a drug regularly they come to depend on it – they are **addicted**. 'Hard' drugs such as heroin and cocaine are particularly addictive. Once the body becomes used to a drug, trying to go without it causes painful withdrawal symptoms. Addiction may alter a person's lifestyle and relationships, often for the worse. Continued abuse of drugs can irreversibly damage the body organs (lungs, kidneys, liver or brain), leading to premature death.

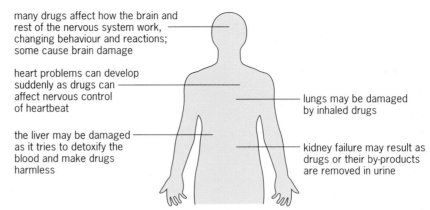

How do drugs affect the body?

Solvent abuse

Almost all of the deaths shown in the graph above could have been prevented – they were the result of abuse. Most deaths were of people under 20 years old; causes included liver and kidney damage, heart failure, suffocating on vomit or plastic bags, and trauma. Of the people who died in 1991, 62% had used solvents before.

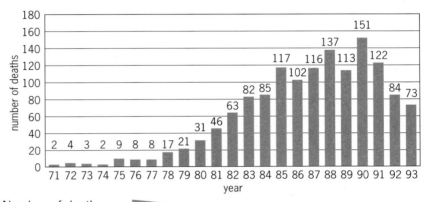

Number of deaths in Britain from sniffing solvents (from St George's Hospital Medical School, with kind permission)

Everyone should be educated about the dangers of drugs and solvents. This is just one of the many leaflets available

★ THINGS TO DO

1. The percentage of solvent abuse deaths due to different causes in one year was:

   - 50% direct poisonous effect on heart and other organs
   - 13% inhalation of vomit
   - 12% accidents while intoxicated or hallucinating
   - 8%   plastic bag suffocation
   - 17% other or unknown causes

   a) Make a pie chart using this data.
   b) What could the 'other causes' be that resulted in 17% of the deaths?

2. a) Discuss in a group why people start taking drugs and why it is hard to stop.
   b) Are there any ways in which the use of drugs could be reduced? If so, how?
   c) Draw a poster to convey your views.

1.28 Everyday harm?

Alcohol

Alcohol changes behaviour, reduces inhibitions and makes some people silly, aggressive or suicidal. Excessive intake of alcohol slows down the brain and impairs coordination. It can lead to unconsciousness, longer-lasting coma or death (e.g. from inhaling vomit). In addition alcohol should not be drunk when certain common medicines are being taken as the combination can harm the body.

Some recent studies, however, suggest that a moderate weekly intake of alcohol may reduce the risk of heart attack. The illustration shows the maximum amounts that are currently thought to be safe to drink.

One in every six road deaths is related to alcohol. Driving really is less safe after drinking alcohol, as it can slow down response time and lessen judgement

Alcohol can be an addictive drug. Heavy drinking over a long period of time can cause damage to the liver, as well as to the heart, kidneys, stomach and brain. The liver absorbs alcohol from the blood, making it harmless, but can itself be damaged in the process. Cirrhosis of the liver, brought on by alcohol abuse, can be life threatening as the liver cells gradually die and are replaced with other cells (scar tissue), which cannot do the same job.

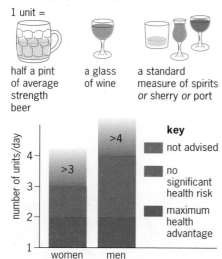

Recommended maximum daily intakes of alcohol for men and women

Smoking

Tobacco contains nicotine, an addictive drug. It also contains harmful chemicals, including tar, which are released as the tobacco burns. Breathing those substances into the lungs can cause irreparable damage both to the lungs, and to other organs, including the heart. Smokers risk damaging not only their own health, but also that of others. 'Passive' smoking (breathing in smoke from someone else smoking) increases a person's risk of developing smoking-related diseases such as lung cancer, emphysema and heart disease.

Smoking is a difficult habit to kick because of the nicotine. Yet giving up is the only way to reduce the risk of smoking-related disease, particularly lung cancer. This disease is hard to detect in its early stages and curable only by removing the affected part of the lungs. Only 25% of patients who have this operation survive for more than 5 years. It is not true to say that you would necessarily develop lung cancer if you smoked. But your chance of getting it would be much higher.

Some facts about smoking

★ THINGS TO DO

1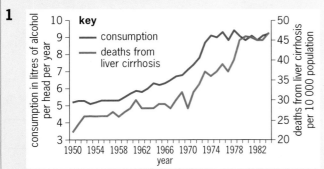

Consumption of alcohol and deaths from liver cirrhosis, in England and Wales (1950–1984) (from *Acquire* no.8, Summer 1992, p.6, Alcohol Concern, with kind permission)

a) Describe the trend in alcohol consumption from 1964 to 1974, and from 1974 to 1984.
b) Describe the trend in deaths from cirrhosis during the same time intervals.
c) What link could there be between alcohol consumption and deaths from liver cirrhosis?
d) One hypothesis is that drinking alcohol causes death from cirrhosis of the liver.
What extra information would you need before deciding whether this hypothesis is true or not? Why would it be difficult to test the effect of excessive alcohol intake using humans?
What are the arguments for and against using animals to test the effect of alcohol abuse on the liver?

2 a) Describe how the death rate varies:
(i) across the country, (ii) by age.
b) To achieve a reduction of deaths from lung cancer, virtually all cases of which are due to smoking, health authorities must consider how to persuade people not to smoke. Possible actions include:

- enforcing the law that prevents the sale of cigarettes to young people,
- increasing teaching in schools about the dangers of smoking,
- encouraging workplaces to become non-smoking areas,
- preventing people from smoking in public places out of doors,
- advertising the dangers of smoking on local TV and radio,
- asking the government to increase taxes paid when buying tobacco.

Discuss the possible benefit of these measures and try to think of other ways to reduce deaths from lung cancer. Write a letter to your local health authority suggesting what could be done.

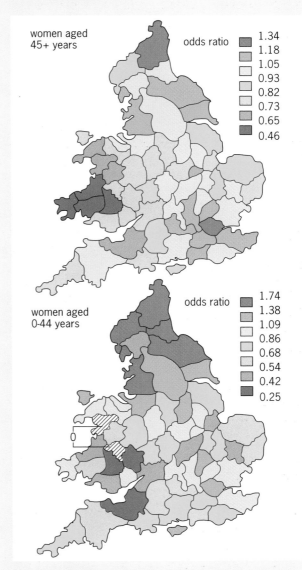

Death rates of women from lung cancer by age, in different counties of England and Wales (from *Guardian Education*, 8 March 1994, p.10, with kind permission)

Exam questions

1 The drawing shows some parts of a human body.
a) What is the name of each of the parts labelled **A**, **B**, **C** and **D**? (4)

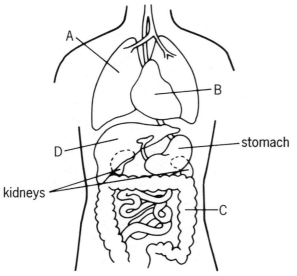

b) Complete each sentence by describing the main job of each of these parts of the body.
The main job of the stomach is ....
The main job of the kidney is .... (2)

c) Choose words from this box to complete the following sentences. (4)

cells   organs   organisms   tissues

.... such as ourselves are made from millions of .... Similar cells work together as.......... .
These work together to form .... , such as our stomach and kidneys.

(SEG, 1995)

2 The diagram shows part of the digestive system. Put the following in the correct boxes to complete the flow diagram below.
A Food is swallowed down the gullet.
B Saliva is mixed with food to soften it.
C Food is digested in the intestine.
D Food is digested in the stomach. (3)

(NEAB, 1995 (part))

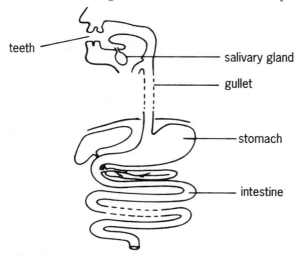

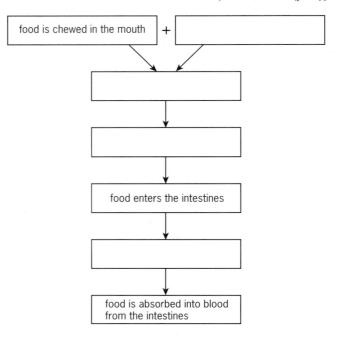

3 The diagram shows an outline of our circulatory system.

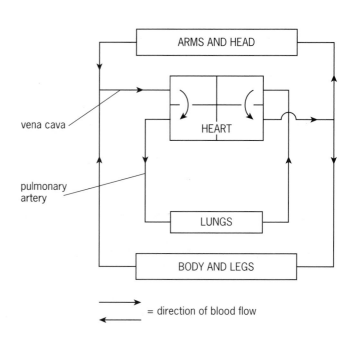

A blood cell travels from an arm to a leg. Fill in the gaps in [a copy of] the flow diagram to show the route it would take through the body. (4)

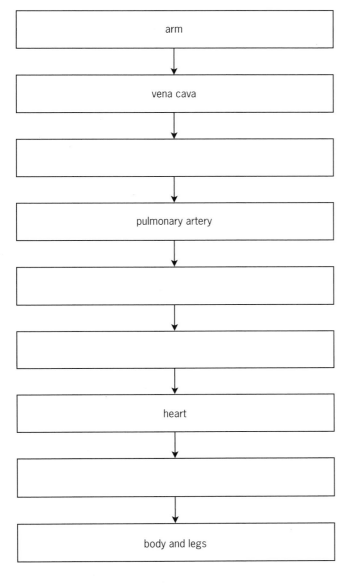

(NEAB, 1995 (part))

4 a) The diagram represents an artificial kidney machine. Kidney machines are used by people whose own kidneys stop working. The machine works by a process called dialysis.

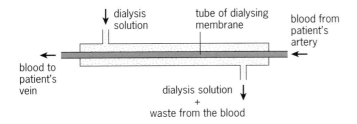

It is important that the dialysis solution contains glucose at the same concentration as it is in the person's blood. Explain why. (3)
(SEG, 1995 (part))

5a) The diagram below shows part of the body seen from the front.
Write down the letter which shows the position of the heart. (1)

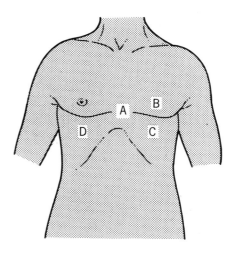

b) The pulses of three girls were measured every minute. Together they walked, then ran and then walked again. The graph below shows how their pulses changed with time.

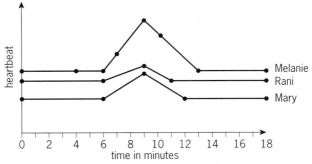

i) For how many minutes did the girls run? (1)
ii) The three girls ran the same distance at the same speed.
Which **one** of the three girls is the fittest?
State **two** reasons for your answer. (3)
(SEG, 1995 (part))

EXAM QUESTIONS

6 The table shows the effect of different amounts of alcohol on a man's body.

Number of alcohol units the man has drunk	Blood alcohol level mg/ 100 cm³	The effect on the body
2	30	feeling good, no cares
4	60	some loss of muscle control
6	90	serious loss of muscle control
14	300	unconscious

a) A man has 75 mg of alcohol per 100 cm³ blood. His body can break down 1 unit of alcohol per hour.
How long will it take to completely break down the alcohol? Show your working. (3)
b) In England it is illegal to drive with more than 80 mg alcohol in 100 cm³ blood.
What effect does this level have on the body? (2)
(NEAB, 1994 (part))

7 a) The conversion of proteins to amino acids can be represented as shown below.

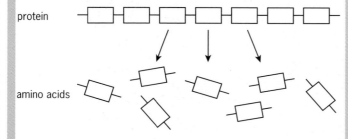

i) What name is given to the conversion of proteins to amino acids in the body? (1)
ii) State **one** use of amino acids in the body. (1)
b) Antibodies are proteins produced by the **immune system** when humans and animals are infected. Some antibody-producing cells can be removed from animals and used to produce large quantities of one type of antibody. These antibodies can be injected into patients to help fight certain diseases.
i) Which part of the immune system produces antibodies? (1)
ii) The effectiveness of a person's immune system can be reduced by the use of drugs. State **one other** way this can happen. (1)
iii) Explain **one** reason why it might be better to use the extracted cells to produce large quantities of one type of antibody rather than large quantities of a mixture of antibodies. (2)
iv) Suggest **one** reason why some people are opposed to the artificial production of antibodies. (1)
c) Protein molecules from other living things which get into the bloodstream can be antigens. Antigens, like antibodies, have positive and negative charges on their surfaces. When antibodies and antigens meet they stick together or "clump". These clumps are broken down and destroyed.
i) Suggest how antibodies might be able to cause antigens to clump together. (1)
ii) After organ transplant operations, patients are given drugs to reduce antibody production. Explain **one** reason why. (2)
(SEG, 1995)

8 The amount of nitrogen combined as compounds in the body is kept constant by a number of processes. Nitrogen is present in cells mostly as amino acids. If amino acids are present in excess they are **deaminated**. This results in the removal of nitrogen, which is excreted as urea.
a) i) What name is given to processes which keep conditions and substances at a constant level in the body? (1)
ii) Why are amino acids necessary in the body? (1)
iii) **Explain** how the body obtains the amount and variety of amino acids that it requires. (4)
iv) Which organ is responsible for the deamination of excess amino acids? (1)
b) The daily excretion of nitrogen compounds from two people is shown in the table below.

Quantity	Person X	Person Y
Urea nitrogen	14.70 g	2.20 g
Uric acid nitrogen	0.18 g	0.09 g
Ammonia nitrogen	0.49 g	0.42 g
Creatinine nitrogen	0.58 g	0.60 g

i) **Explain**, with reasons, how the diet of person **X** differs from that of person **Y**. (2)
ii) If anabolic steroid drugs are taken, the nitrogen balance in the body is disturbed. As a result, the intake of nitrogen compounds greatly exceeds their excretion. What effect could taking these drugs have on the body? (1)
(SEG, 1994)

2
PLANT LIFE

2.1 Growing for gold

Jack Anderson spends hours and hours each year growing perfect plants, which he hopes will win competitions.

Jack explains the stages of growing his leeks to perfection (see diagram).

Plants growing in the wild rarely reach perfection, but show the same features:

- **roots** to anchor them firmly in the ground, and to absorb water and nutrients,
- **stems** to hold up leaves and flowers, and to pass substances up and down,
- **leaves** to absorb energy from sunlight so food can be made by photosynthesis,
- **flowers** where seeds are made during reproduction.

The plant cell

Each part of the plant has its job to do, so that the plant stays healthy and grows well. Plants are like animals – they contain millions of microscopic cells (see Topic 1.1). There are many different plant cells, each doing a different job. Cells in plants have several differences to those in animals, as the diagram shows.

The basic plant cell (parts marked* are common to both plant and animal cells)

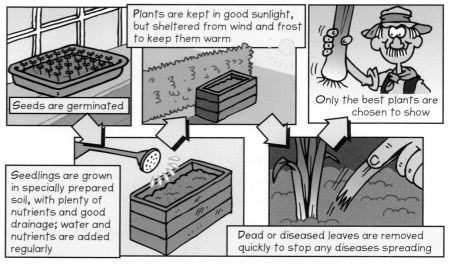
How to raise leeks

The features of a plant

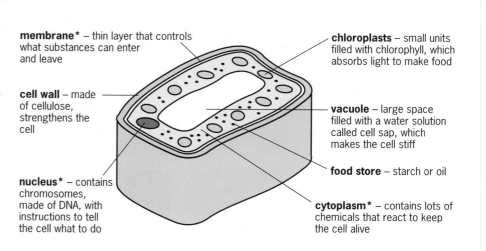

membrane* – thin layer that controls what substances can enter and leave

cell wall – made of cellulose, strengthens the cell

nucleus* – contains chromosomes, made of DNA, with instructions to tell the cell what to do

chloroplasts – small units filled with chlorophyll, which absorbs light to make food

vacuole – large space filled with a water solution called cell sap, which makes the cell stiff

food store – starch or oil

cytoplasm* – contains lots of chemicals that react to keep the cell alive

PLANT LIFE

Plant organisation

Plant cells, like animal cells, are grouped together as tissues (e.g. xylem tissue).

Tissues are grouped together as organs (e.g. roots, stems, leaves).

Organs in plants are much simpler than organs in animals.

Some plants (e.g. simple algae) consist of only one cell.

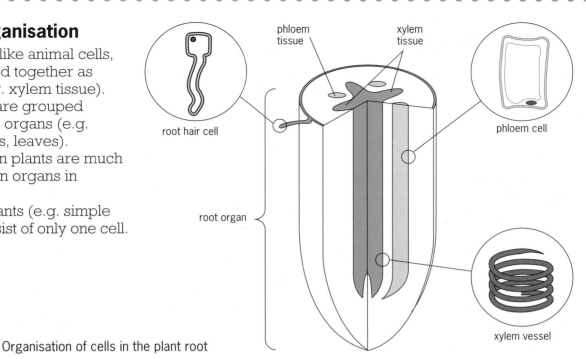

Organisation of cells in the plant root

★ THINGS TO DO

1 a) List the conditions needed by a plant to grow well.
b) Plan a competition to grow the best plants (e.g. radish, or lettuce). Think of rules that everyone must follow. Decide how and when your plants will be judged. Then grow your plants.
c) Each week, until the judging, write a report on your plants' growth.

2 Copy this table and compare the parts of a plant cell with the parts of an animal cell (see Topic 1.1). Some comparisons have been made to guide you.

	Plant cell	Animal cell
nucleus	present	present
membrane		
cell wall		
cytoplasm		
vacuole	large, permanent	small, temporary
food store		glycogen, fat
chloroplasts		

3 If possible look at plant cells under a microscope. (If you have not used a microscope before, make sure your teacher or technician shows you how.) You might look at ready-prepared sections of parts of a plant, or make some of your own slides and look at them using a microscope. You could, for example:

- Peel away the 'skin' of an onion bulb, and put it on a drop of water on a microscope slide. Flatten with a cover slip, then look at it under a microscope.
- Repeat the steps above, but using a drop of iodine solution instead of water (to stain the cell and make parts easier to see).
- Lay a single leaf from a moss plant on a drop of water on a slide, and flatten with a cover slip.

For each, draw and label what you see. Include the magnification used; calculate the size of the cells.

2.2 A good start

It is spring and Carol is choosing plants for her garden. She wants a border of plants, with small ones at the front and bigger ones behind. She will grow annuals from seed, and buy others as mature plants. Her seed catalogue has information about different plant types.

Cornflower – easy to grow, sow seeds Mar.–May, flowers June–Sept., height 60 cm (A)

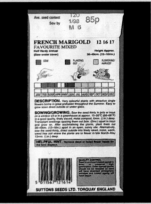

Marigold – not hardy, sow seeds indoors Feb.–Apr., plant out when frosts have finished, flowers June onwards, height 15 cm (A)

Sweet William – sow seeds May–June, flowers June–July, height 45 cm (B)

Delphinium – sow seeds May–July, flowers following June onwards, height 100 cm (P)

The codes A, B, P, which appear on some seed packets relate to the life cycles of the plant:

- A = annual (1 year). The plant grows from seed, flowers, then dies all in 1 year. The next generation of plants must be grown from seed.
- B = biennial (2 years). In the first year, the plant grows from seed. In the second it grows further, flowers and dies. The next generation of plants grows from seed.
- P = perennial (many years). In the first year the plant grows from seed, and may flower if sown early. In the second it grows further and flowers, but does not die. The part of the plant above ground may die in the winter, but the part that is underground survives. In spring it once again grows into a mature flowering plant. This cycle repeats every year.

PLANT LIFE

Old potatoes, new potatoes

Potatoes are perennial plants that are easy to grow from 'seed' potatoes. These are underground perennating organs (**tubers**) – parts of the potato plant which survive the winter after the rest of the plant has died (see Topic 4.7).

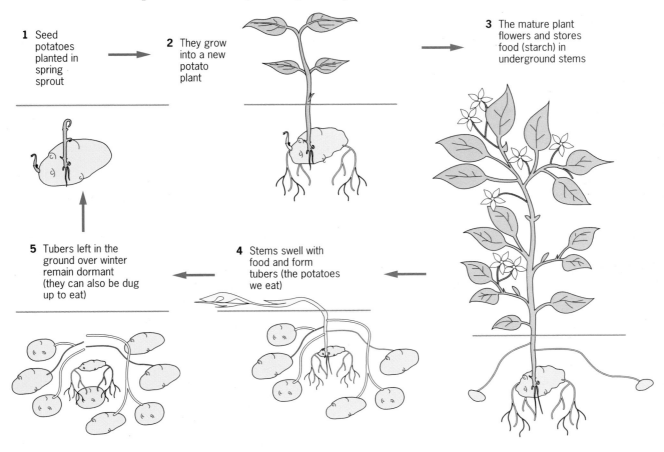

Life cycle of the potato

Seeds

Seeds can be collected from mature plants and stored. The next spring they can be planted for the following year's crop.

When planted, the combination of moisture and warmth makes seeds **germinate** – they begin to grow into new plants. All seeds have a protective **seed coat**, with an **embryo** plant inside, and a store of food called the **endosperm** (see Topic 4.9), or **cotyledon**. The amount of food stored can vary. The diagram shows the inside of a seed.

The seeds in a packet will be different sizes. The larger ones (i.e. those with more endosperm) will have the best chance of growing well. This is because the more food the embryo plant has the more it can grow in the early stages of its life, leading to stronger seedlings.

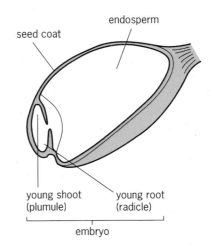

A seed cross-section, showing the embryo inside it (wheat seed)

Germination

Germination is the growth of seeds into young plants called seedlings. It can be thought of as a series of stages.

Germination success

Seed packets always include some details about the quality of the seeds and how they should be germinated. They may say something like:

> These seeds are tested in our laboratories to ensure purity and the highest germination success. Standard seeds. EU rules and standards.

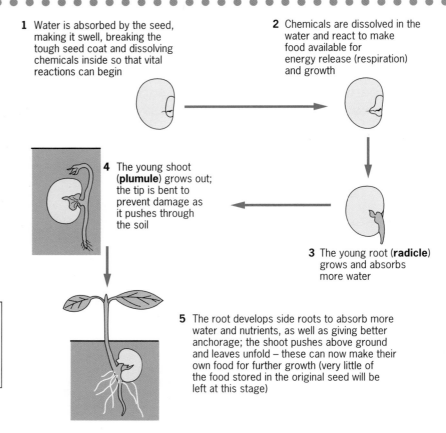

Stages of bean germination

One method to test seeds is a percentage germination test, that is, how many out of 100 germinate. This can be done with a 'germination tester' – a plastic tray with 100 small wells in it. A seed is placed into each well, and the tray is left in conditions that encourage germination. The results of one test are shown in the diagram. Seeds in tray 1 would pass the test, so other seeds from the same batch can be sold. But seeds in tray 2 would fail; if further tests gave the same results this batch of seeds would not be sold.

The seeds of most species of plant can remain in a **dormant** state (i.e. not growing) for many weeks to many years. The embryo plant inside the seed uses a minimum of energy during dormancy, saving its food store until the conditions for germination are right. Seed packets normally show the date when they were packed and usually give an expiry date. The longer seeds are stored, the less likely they are to germinate, as more and more gradually die or become diseased.

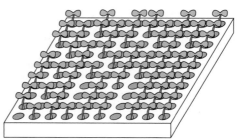

Germination tests

★ THINGS TO DO

1 a) Use a seed catalogue to find out the names of some plants that could:
(i) grow in shade, (ii) need bright light, (iii) grow to over 50 cm tall, (iv) grow smaller than 25 cm, (v) grow in heavy soil, (vi) grow in light, sandy soil.
b) Explain why the plants grown from a packet of seeds will not always flower at the same time.
c) What are the advantages and disadvantages of growing perennial plants?

2 a) Traditional growers often chop seed potatoes up so that they get two or three smaller seed potatoes, all of which will grow into mature plants. What are the advantages and disadvantages of doing this?
b) What relationship could there be between the size of a seed potato and how well the new plant grows? Write a plan describing how to investigate your prediction.
c) If possible, test your ideas. Write a report describing what you find.

3 Ismail thought that 'seeds need water, air and warmth to germinate'. He did this investigation:

Tube	Conditions		
	Water	Temperature	Air
1	yes	warm	yes
2	yes	cold	yes
3	yes	warm	no
4	no	warm	yes

The results are shown in the diagram.

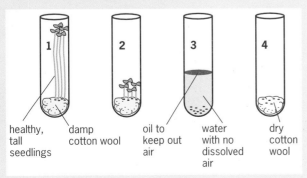

a) Do the results support his hypothesis? How?
b) He covered each tube with foil to stop water loss and to keep the seeds in the dark. Why was this important?
c) How would he make his test fair?

4 A trial germination was carried out on radish seeds from packets that were out of date. The percentage germination rate was determined. The results were:

Years after expiry date	% seeds germinating
1	58
2	30
3	15
4	9
6	6
7	5

a) Plot the data on a line graph.
b) Explain what the graph suggests about germination success.
c) What germination rate would you expect for seeds 5 years over their expiry date?
d) What germination rate would you expect for new seeds?

5 A gardener suggested that the best way to germinate sweet pea seeds was to: (1) scratch or cut the surface of the seed, and (2) soak the seeds for up to 24 hours in warm water before sowing.
   Plan and carry out an investigation to test these ideas. Write a report. You could send your findings to a seed supplier and ask for their comments.
   You could also try to find:
- the best temperature to soak sweet pea seeds before sowing,
- if sweet pea seeds can be soaked for less than 24 hours and still germinate well.

2.3 Measuring growth

What is growth?

If you need to keep cutting (mowing) a lawn it means that the grass is growing. The amount of grass cuttings collected shows how much the grass has grown. **Growth** is a permanent increase in size. How quickly this happens, the *rate* of growth, varies according to the seasons and the weather. Grass grows faster in spring and after rain.

Plants grow buds, which then grow into flowers

The leaves of a deciduous tree grow early in the year, then die and fall off in autumn

Trees grow wider. It takes time, but leaves a lasting impression

The cell cycle

Plants, like animals, are made of millions of microscopic cells. Plants grow as:

- cells become bigger,
- new cells are made.

Most cells go through a cycle of activities during growth.

Growth spurts

The growth rate of plants (as in animals) is controlled by plant hormones (see Topic 2.14). Changes in the amount of food, water, temperature and light also affect how quickly a plant grows.

A change in growth rate can take place over several weeks, or be fairly sudden. Deciduous trees (e.g. oak, sycamore) grow quickly in spring and summer, but hardly at all in winter. When a deciduous tree trunk is cut, a pattern of **annual rings** is seen. These show the different amounts of growth from season to season. The number of annual rings equals the age of the tree. Wider annual rings result from better growth, when conditions are good. Studying the rings of older trees gives important clues about long-term weather patterns.

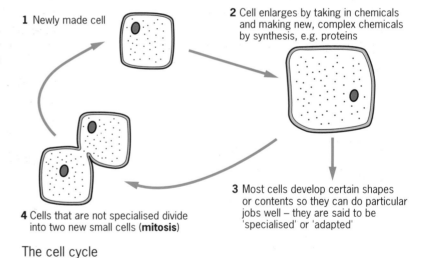

The cell cycle:
1. Newly made cell
2. Cell enlarges by taking in chemicals and making new, complex chemicals by synthesis, e.g. proteins
3. Most cells develop certain shapes or contents so they can do particular jobs well – they are said to be 'specialised' or 'adapted'
4. Cells that are not specialised divide into two new small cells (**mitosis**)

Annual rings on cut trees

PLANT LIFE

You may have noticed how grass covered by tents becomes yellow after a few days. This effect is called **etiolation**. It happens because plants grow much quicker, becoming tall and thin, when shaded from the light. They respond to a lack of light by growing more to search for the light they need to survive. Similarly, some plants tend to grow more quickly at night than in the day.

Etiolated plants and normal plants

★ THINGS TO DO

1 Sprouting beans, like mung beans, are delicious to eat. They grow quickly if kept in a clean container and rinsed with water twice daily.
   a) How could you use a ruler or balance to measure their growth?
   b) What factors might affect their rate of growth?
   c) Plan an investigation of how quickly they grow, and carry it out.

2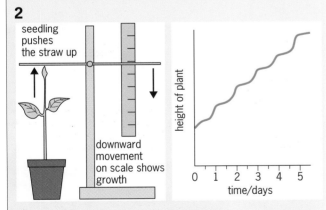

The growth of a potted seedling can be measured by using a pivoted straw, which is pushed up by the seedling as it grows (see diagram above). The distance moved on the scale by the other end of the straw represents growth. The graph shows the results of using this technique.
   a) Describe how the plant has grown over 5 days.
   b) Realistically, how often could measurements be made?
   c) Computer data-logging equipment can be used to measure growth automatically. Find out how it works
   d) If possible, use this type of equipment to investigate how different variables can affect the growth of a plant over several days.

3 Foresters can tell the age of coniferous trees by measuring their girth (distance around the trunk). These trees grow steadily each year, as the data show.

Age of tree/years	Girth/cm
10	46
20	90
30	135
40	176
50	222

   a) Draw a line graph of the data.
   b) What girth would you expect a 35-year-old tree to have?
   c) Explain the reasoning behind your estimate.
   d) Trees like oak grow quicker in summer than in winter. The size of 'annual rings' on the cut trunk will show the amount of growth each year. Why don't foresters age oak trees by measuring their width?

2.4 Plants are producers

Plants do not take in food ready-made like animals. Instead, they make their own food – they are called **producers**. The series of chemical reactions by which a plant makes its food is called **photosynthesis**.

How do plants grow?

People used to believe that plants grow because they absorb soil through their roots. That idea was shown to be wrong by Jan Baptista van Helmont in 1648. He suggested that the tree might instead be absorbing a gas from the air, which it then used to make food. We now know that plants do indeed absorb a gas (carbon dioxide) from the air during the day, and give out another (the oxygen which other living things need to survive), during the process of photosynthesis.

Photosynthesis is a series of chemical reactions that are summarised in the diagram.

Water and carbon dioxide are needed for photosynthesis; carbohydrate and oxygen are the products. During photosynthesis some of the light energy falling on the plant is absorbed, and converted into chemical energy and stored as carbohydrate molecules.

The water needed for photosynthesis enters through the roots, and is carried up to the leaves by xylem cells in the stem (see Topic 2.1). Carbon dioxide is absorbed through small pores in the leaves.

Photosynthesis occurs only in the light, which provides the energy needed for the chemical reactions. Light is absorbed by coloured chemicals called pigments, the commonest being **chlorophyll**. It makes plants look green because it reflects green light (i.e. it absorbs all other colours).

Chlorophyll is packed into oval units called **chloroplasts** inside the stems and leaves in the cytoplasm, especially in the palisade cells (see photo). The more chloroplasts in a cell, the more light can be absorbed.

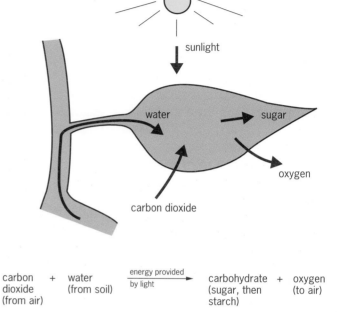

carbon dioxide (from air) + water (from soil) $\xrightarrow{\text{energy provided by light}}$ carbohydrate (sugar, then starch) + oxygen (to air)

Photosynthesis in the leaf

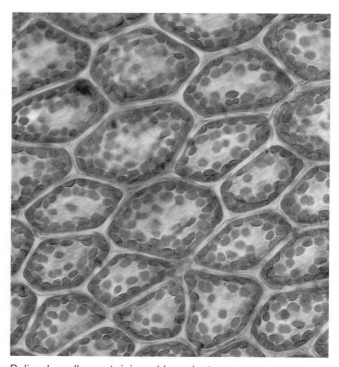

Palisade cells containing chloroplasts

PLANT LIFE

Sugar, such as glucose, is the first type of carbohydrate made. This, if not needed immediately by the plant, is changed into starch in the leaf cells. Starch is insoluble and can be stored for later use. When required the starch is changed back into soluble sugar and passes through the phloem to the part of the plant where it is needed. It is used or stored in other organs (see Topic 2.5).

The leaf therefore acts as a food-making factory and supplies the whole plant with food (carbohydrate). Minerals like nitrogen are also involved in making the wide range of chemicals needed by the plant to grow, stay healthy and reproduce (see Topic 2.7).

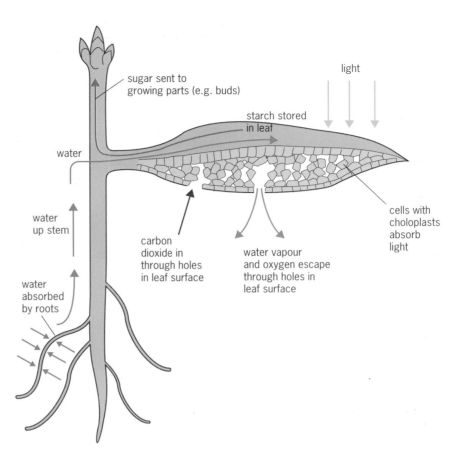

Photosynthetic pathways in a plant

Lack of daylight in winter slows down plant growth, even in the greenhouse. Our Growlights can add the light that is missing, even in the darkest days of winter. Now you can even grow plants in converted garages, or cellars!

Nature's energy

Sunlight is a natural source of the energy needed by plants to make their food. Without it, photosynthesis cannot take place. Hence most of the food supplies are made during summer when the long days and higher temperatures provide ideal conditions. In winter months the short, cold days mean that plants cannot get enough energy to produce the food they need. Artificial sources of light such as 'Growlights' are used by many market gardeners. They provide light for 24 hours a day, all year. This means that food crops can now be grown throughout the year, even when they would formerly have been in short supply. However, the artificial light must contain enough red and blue light in it because these are the wavelengths most readily absorbed by chlorophyll.

PLANTS ARE PRODUCERS

Maximising light

The leaves of a plant are adapted to absorb as much light as possible. Tall plants, for example, have the advantage that their leaves are not shaded by others – their leaves are held out by the stems so that most of their surface is exposed to light.

Plants that grow under the trees in thick woodland often begin growing in early spring, before they are shaded by the tree leaves above (see Topic 3.3). This enables them to photosynthesise and make their food, which they store below ground in their bulbs, to keep them supplied until the following spring.

Leaves rarely overlap. They are arranged as a 'mosaic' so that one leaf absorbs light that bypasses others

★ THINGS TO DO

1.

Month	Average daily hours of sunshine
J	1
F	2
M	3
A	5
M	6
J	7
J	6
A	6
S	5
O	3
N	2
D	1

Average sunlight hours over 12 months in London

a) Use the information in the table to draw a graph showing how the number of hours of daylight changes throughout the year.
b) Describe how the rate at which photosynthesis takes place will change throughout the year. Draw three lines on one graph, showing how the rate of photosynthesis will change in a 24-hour period: (i) in winter; (ii) in spring; (iii) in midsummer.
c) If possible, use a light meter and data logger to measure light levels automatically over 24 hours. Use your results to describe how the rate of photosynthesis will change during the 24 hours.
d) i) Devise a way of investigating the effect of different colours of light on plant growth. Plan and carry out this long-term investigation.
ii) Explain why the colour of light could have any effect.

PLANT LIFE

2 Leaves can be crushed and tested for sugar using either Benedict's test or indicator papers. A test for starch is shown in the diagram.

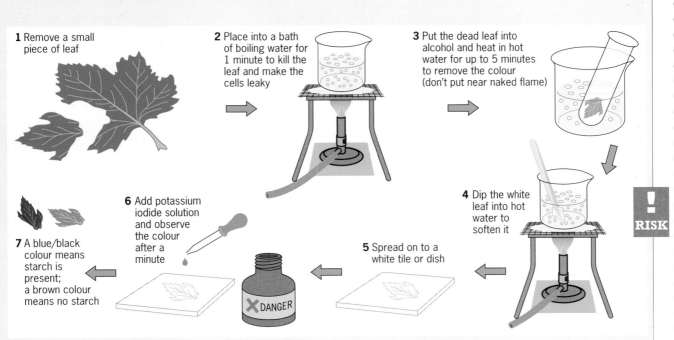

Testing a leaf for starch

A test was made of the leaves from potted plants that had been treated in a variety of ways to find out which of them contained starch.

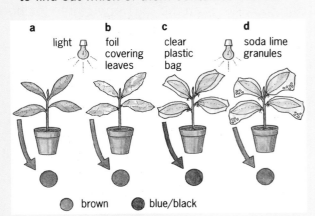

Colours after the starch test: **a** plant left in darkness for 2 days; **b** plant left in darkness for 2 days, then leaf covered in foil for 2 days; **c** plant left in darkness for 2 days, then leaf covered in clear plastic in light; **d** plant left in darkness for 2 days, then leaf covered in clear plastic containing soda lime, in light (soda lime absorbs CO_2)

It was concluded that a leaf loses starch in the dark, but makes starch in the light if it has a supply of carbon dioxide.

a) Why do you think a leaf that had been left in darkness for 2 days was tested?
b) Use these results to explain how you could tell that both light and carbon dioxide are needed to make starch.
c) If possible, repeat the experiment, preferably with a number of different plants.
d) Use a variegated plant like the one in the photograph to test if chlorophyll is needed for making starch.

Leaves of *Coleus*

2.5 Food factories

Photosynthesis can take place in any part of a plant that contains chlorophyll (that is, the green parts). But up to 95% of the food is made by photosynthesis in the leaves. This is because leaves are adapted to make food more efficiently than other parts. They have a large surface area to collect lots of sunlight, and most cells are packed with chloroplasts to maximise light absorption. The most important features of the leaf are shown in the diagram.

Other plants' leaves have slight differences that help them survive.

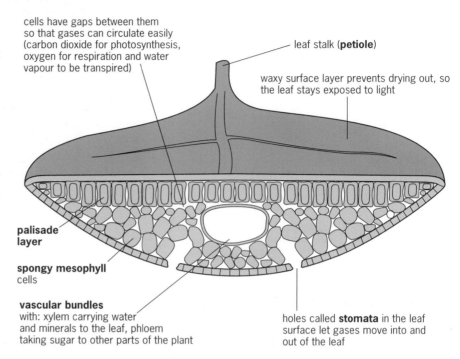

The main features of a leaf

Food on the move

During photosynthesis, carbon dioxide combines with water to form carbohydrates (see Topic 2.4). If the plant is exposed to radioactive carbon dioxide, then the carbohydrates formed are radioactive. By tracing the path of the radioactive carbohydrates through the leaf, we can see where they go and how they are used. The photograph was made using photographic film that forms an image of the radioactive compounds inside the plant. The darkest parts of the image show where most of the radioactive carbohydrates are.

These experiments have shown that sugar made by photosynthesis is used by a plant in many different ways.

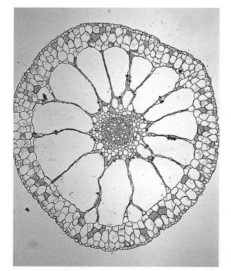

Leaves of water plants have many air gaps, which give them extra buoyancy, helping them to float close to the water surface where there is more light. Here they can absorb more energy for photosynthesis (a waxy surface layer is not needed since the leaf is surrounded by water)

Leaves blackened by radioactive carbon

PLANT LIFE

Using food

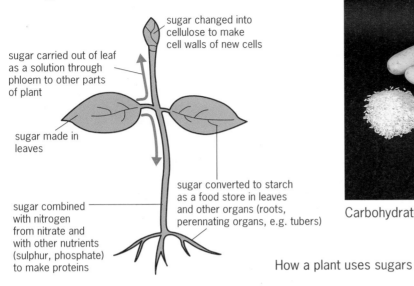

How a plant uses sugars

Carbohydrate foods

★ THINGS TO DO

1 Prepare your own thin sections of leaves on microscope slides using the instructions below. You could try the leaves from a plant such as privet. **RISK**

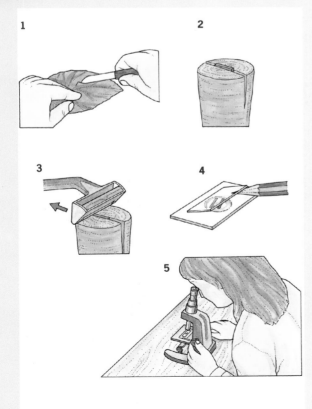

2 An experiment was carried out on a potato plant with one leaf exposed to radioactive carbon dioxide (^{14}C). The amount of radioactivity in different parts of the plant was measured after 24 and 48 hours.

Part of plant	Average amount of radioactivity/c.p.m.	
	After 24 hours	After 48 hours
leaf exposed to ^{14}C	989	476
leaf stalk	354	132
flower bud	56	367
roots	69	78
tuber	74	145

There was an average background level of radioactivity around the plant of 35 counts per minute (c.p.m.) after 24 and 48 hours.
a) Why should the background levels of radioactivity be measured?
b) Why is the leaf so radioactive after 24 hours?
c) Why is it less radioactive after 48 hours?
d) Describe how the levels of radioactivity vary in different parts of the plant. Explain why the levels vary like this.

2.6 More and more

The rate at which photosynthesis takes place is important in many ways. To a farmer a good summer means one that provides enough rain for his crops, but accompanied by warm, clear days with plenty of sunshine. Under these conditions, food crops grow well. This means that there will be a good supply of the food we need at reasonable prices. Photosynthesis in water plants provides oxygen, which pond-living animals need. Poor conditions can lead to a lack of oxygen and the death of the animals that live there.

What affects the rate of photosynthesis?

The rate at which chemical reactions take place depends on:

- the *amount* and *concentration* of the reactants – generally, the more concentrated the reactants the faster the reaction takes place,
- the *temperature* – generally, the higher the temperature the faster the reaction takes place.

Because photosynthesis is a series of chemical reactions, the rate at which plants can photosynthesise will depend on the amount of raw materials available and the temperature – but only up to a point.

As you can see from the left-hand graph (below), increasing the amount of light increases the rate at which photosynthesis takes place, but it reaches a stage where further increases in the amount of light have no further effect.

This may be because:

- the leaves are unable to absorb any more light than they are already absorbing,

A cereal harvest

- the temperature is limiting the reaction rate,
- the leaf cannot absorb carbon dioxide any faster than the current rate,
- water cannot be made available any faster,
- or any combination of these factors.

In other words, the rate of photosynthesis is limited by other factors (called *limiting factors*). The amounts of light, carbon dioxide (centre graph) and temperature (right-hand graph) are all limiting factors. The overall rate of photosynthesis is controlled by the availability of each.

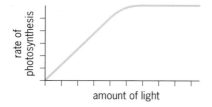

Rate of photosynthesis versus light

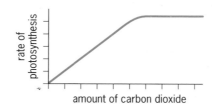

Rate of photosynthesis versus carbon dioxide

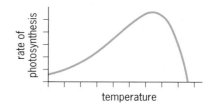

Rate of photosynthesis versus temperature

PLANT LIFE

★ THINGS TO DO

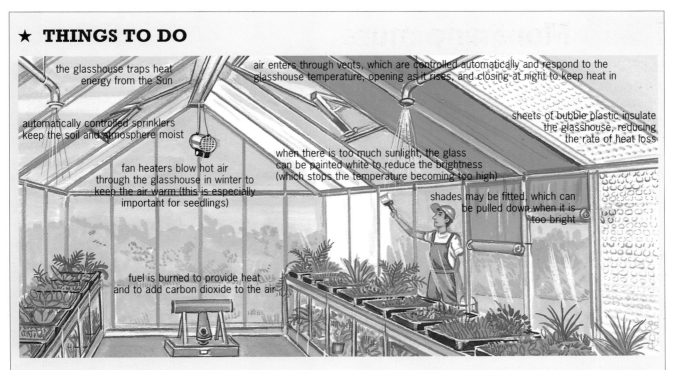

1 Some of our food comes from plants grown under glass. Glasshouses allow the control of factors that affect the rate of photosynthesis (and therefore growth). Explain how each of the features of the greenhouse described in the diagram helps plants to make food more effectively.

2 Matt carried out an investigation to find out whether the water plants in his aquarium produced more oxygen when the light was on. When the lamp was 10 cm from the aquarium he counted the number of bubbles released in 5 minutes. He did this three times to get an average. Then he placed the lamp progressively further away and counted the bubbles each time. These are his results.

Distance of lamp from plant/cm	Bubbles released in 5 min	Average
10	127, 70, 100	
20	75, 69, 68	
40	36, 48, 30	
60	18, 21, 24	
80	14, 14, 12	

a) Copy and complete his table of results.
b) Draw a graph of them. What does it tell you?
c) What steps would Matt need to have taken to ensure his test was fair?
d) It was suggested that the increased number of bubbles could be due to a temperature increase in the water when the light was on, rather than the extra light falling on the plant. How could Matt check which idea was right?
e) If possible, repeat this experiment, with any improvements you wish (you could use a light meter); compare your findings with Matt's.

3 Find out about the greenhouse effect (see Topic 3.14). Write an account of what could happen to plant growth in Britain with increased global warming. Include both possible benefits and possible drawbacks. Explain why they may, or may not, happen.

2.7 Feeding plants

Look along the shelves in a garden centre and you'll find plenty of 'plant foods'. The expression 'plant food', though often used, is wrong scientifically, however, as plants make their own food by photosynthesis (see Topic 2.4). First they make sugar, a type of carbohydrate, which is then changed into whichever different types of food the plant needs (such as protein and oils). For this, they need simple chemicals, which are absorbed (in solution) from the soil. These are called minerals, or nutrients, and are what gardeners call 'plant foods'.

Gardeners spread fertiliser on their lawns and gardens in spring and autumn. Fertilisers contain compounds of the elements that plants need for healthy growth. The main fertilisers contain compounds of nitrogen (N), phosphorus (P) and potassium (K). Other elements are also included in much smaller quantities. After a few days the lawn will look greener and healthier as the grass plants absorb and use the nutrients provided

Growing without minerals

Unhealthy looking plants may be diseased. However, more often the cause is a lack (deficiency) of minerals. Deficiencies cause particular changes in the plant's appearance, as shown in the photographs.

If a deficiency is recognised early, the missing mineral can be added to the soil in which the plant is growing. It can be watered on (as a solution) or spread on the soil as a powder. Once absorbed, the mineral is used quickly and growth returns to normal.

a oilseed rape   **b** grapevine   **c** tomato   **d** potato

Mineral deficiency conditions: **a** lack of nitrogen (from nitrate); **b** lack of potassium (from potash); **c** lack of phosphorus (from phosphate); **d** lack of magnesium

A special trick

Young people are fascinated by a plant called a Venus fly trap. This plant is carnivorous. Normally it grows in poor soil that lacks the nitrate needed by the plant to make protein for growth. So it obtains nutrients by trapping and digesting small animals. When a small animal touches the leaves they close, trapping it. The animal dies and its protein is digested to release nitrate, which is then absorbed by the leaves.

The bog plant called the Sundew is also carnivorous. It uses a different method to trap tiny animals, producing a sticky liquid that acts like glue on its leaves. Once again, the animals are digested and the nitrate-rich liquid absorbed.

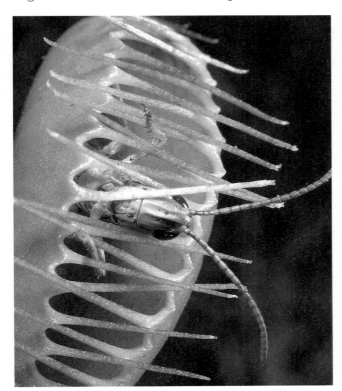

Venus fly trap plant

A damsel fly trapped on a Sundew plant

Absorbing and using minerals

Minerals occur in most soils as rocks are broken down into smaller and smaller particles. The minerals dissolve in water, forming a solution that is absorbed either through the roots of plants growing in soil or, in the case of aquatic plants growing in ponds and streams, by the plant as a whole.

Roots are designed to absorb water, and anything dissolved in it, as effectively as possible. The minerals in solution pass into root hairs, and then to the xylem in the middle of the root (see Topic 2.1). Solutions of minerals are absorbed into root hairs by two processes: diffusion and active transport (see Topic 2.13). The solution then travels up through the stem in both the xylem and the phloem tissues to other areas.

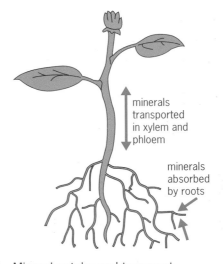

Mineral uptake and transport

FEEDING PLANTS

these minerals are used in large amounts:

nitrate to make protein,

potassium to help enzymes work in respiration and photosynthesis,

phosphate, needed in the reactions of respiration and photosynthesis

other minerals needed in smaller amounts are:

sulphur to make protein,

calcium to make cell walls,

magnesium to make chlorophyll

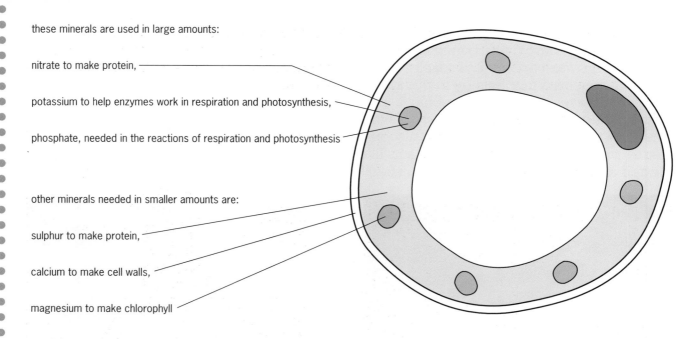

Using minerals within a cell

★ THINGS TO DO

1 The table shows the results of some tests done by *Which?* magazine using different tomato fertilisers. The relative amounts of the main nutrients are given as 'NPK values', where N represents nitrate, P represents phosphate, and K represents potassium.

Use the information in the table to answer these questions in your notebook.
a) Which of the fertilisers was the cheapest to buy over a whole season?
b) Which of the fertilisers were solids which would be dissolved in water before use?
c) Which nutrient had a higher percentage than the others? Why do you think manufacturers make their fertilisers like this?
d) Which fertiliser had the largest crop of tomatoes after: (i) 8 weeks (ii) 11 weeks?

A comparison of tomato fertilisers (data from *Which?*, with kind permission)

Tomato fertiliser	Size	Price /p	Nutrients			Cost of fertiliser used over 1 season /p plant^{-1}	Weight of tomatoes/g		
			N	P	K		after 4 weeks	after 8 weeks	after 11 weeks
Bio	440 ml	79	6	5	9	47	2.55	5.14	7.38
Boots	500 ml	49	4	4	7	28	1.93	4.82	6.64
Tomorite	440 ml	85	4	4.5	8	96	2.75	6.05	8.88
Tomato Plus	440 ml	83	4	4	7	70	2.27	5.19	6.04
Maxicrop	500 ml	78	5.1	5.1	6.7	30	2.16	4.40	5.79
Phostrogen	243 g	37	10	10	27	17	2.66	5.05	5.95

e) Draw a bar chart showing the weight throughout the season of tomatoes obtained using each of the fertilisers. How would the bar chart help gardeners to decide which fertiliser was best?

f) Imagine that you were asked to repeat the tests. Write a plan saying what you would have to do. Remember to include things that you would do to make your tests 'fair'.

2

Different types of fertilisers contain different amounts of nitrate, phosphate and potassium. The table shows NPK values for different uses.

Fertiliser	NPK value	Use
Springlawn	14:3:7	spring/summer lawns
Nomow food	3:8:4	autumn lawns
Plantfood	10:10:27	general purpose

a) Why is the term lawn 'food' not strictly correct?
b) Describe the similarities and differences between the different fertilisers.
c) Why should more nitrate be available for grass plants over the summer months?
d) Feeding a lawn in autumn encourages grass roots to grow strongly. Which mineral is most needed for this?
e) What other minerals might you expect to find in a fertiliser? If possible, look at some packets to see what they contain.

3 Young plants such as peas can be grown successfully in containers filled with a solution of minerals. A simple way to do this is shown in the diagram.

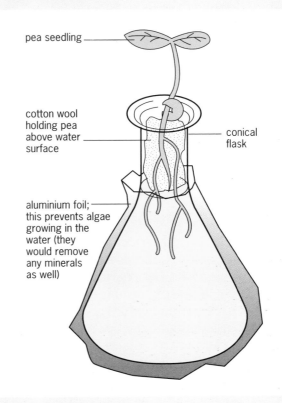

a) How could you use pure water to show that the pea plants need minerals to grow normally?
b) Why is it better to use young plants rather than older ones?
c) It is possible to make solutions in which certain minerals are missing. You could, for example, prepare a solution that contains all the minerals a plant needs except nitrate. How could you use this solution to investigate the effect of nitrate deficiency on plant growth?
d) If possible, test the effect of growing young plants without different minerals. Prepare an article for a gardening magazine describing what you find out. You could add pictures.

2.8 Respiration matters

Living things need energy to stay alive. The more active they are, the more energy they use. So the more a plant grows, the more energy it needs.

Plants, like animals, release energy from food by the chemical process of **respiration**. This takes place within cells in the mitochondria (see Topic 1.2). As in animals, most energy is released when sugar reacts with oxygen to form carbon dioxide, water and ATP.

$$C_6H_{12}O_6 + 6O_2 \rightarrow 6CO_2 + 6H_2O + ATP$$
(sugar) (energy store)

Energy in ATP is converted into forms of energy that the plant can use more easily for growth. Some of the energy formed during respiration also warms up the plant. Carbon dioxide is a poisonous waste product; any not used in photosynthesis escapes from the plant. Respiration takes place throughout both night and day. Because oxygen is used up, it is aerobic respiration (see Topic 1.9).

The respiration rate increases when there is:

- plenty of food,
- a good supply of oxygen,
- warm conditions.

These marrows have been planted on a mound of compost. As the vegetation decays heat is released. The warm conditions increase the rate of respiration in the marrow plants – so they grow quicker

Carbon dioxide made by respiration is released only if it is not used up by photosynthesis. The graph shows how this is related to light intensity during the day.

Not enough air

Most plants grow badly when their roots don't have enough oxygen for respiration. If fields flood after heavy rain, plants can die because the water pushes air out of the soil (it is 'waterlogged'). But other plants are adapted to grow in waterlogged conditions, as their roots can respire without oxygen.

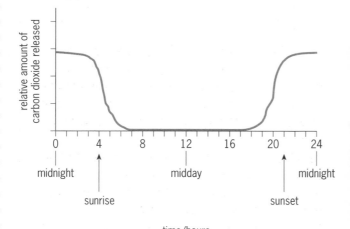

Carbon dioxide release over a day

Rice plants grow well in flooded fields

PLANT LIFE

Large seeds such as peas and beans can also respire anaerobically. So they germinate successfully, even though little oxygen gets to the embryo inside. They can release energy from food without using oxygen by a type of anaerobic respiration called **fermentation**. It can be summarised as:

sugar → carbon dioxide + ethanol + energy released

Anaerobic respiration releases much less energy from food than does aerobic respiration. The energy released is still stored in ATP and is used in the same way regardless of the form of respiration taking place.

Fermentation is used to make beer, wines and alcoholic spirits; yeasts (types of fungi) are used to ferment the sugar, changing it into alcohol

★ THINGS TO DO

1 a) What conditions are needed for aerobic respiration?
b) What conditions are needed for anaerobic respiration?
c) Why do most plants grow slowly if they don't have enough oxygen?

2

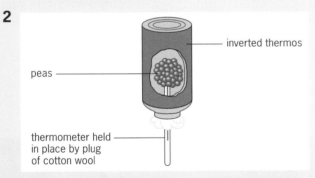

The apparatus shown allows us to monitor the heat energy released by respiring peas. After 24 hours their temperature was 35 °C.
a) Why is a thermos flask used?
b) How will it help to leave some air in the flask?
c) Why should the flask be upside down?
d) The room temperature after 24 hours was 18 °C. Why is the temperature in the flask higher than the room temperature?
e) Someone suggested that heat was produced because the peas were dead and were decaying. How could you determine whether the heat was from respiration or decay?

f) If possible, plan an investigation to record hourly temperature change with a data logger. Plot a graph showing the temperature variation. Explain any patterns in your results.
g) How do you think the temperature in the flask will vary with the number of peas used? Suggest a reason for your prediction. Carry out an investigation to test your idea.

3

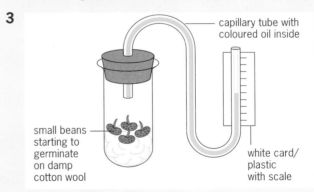

The rate of respiration of germinating beans can be measured by using a container and capillary tubing as shown. The oil drop moves towards the far end of the tube if more carbon dioxide is made by respiration than oxygen is used up.
a) Why is this method suitable for measuring the rate of anaerobic respiration?
b) How could you use it to study the effect of temperature on respiration rate?
c) If possible, try out your plan and explain fully the results you obtain.

2.9 Water for life

Plants, like animals, need water to stay alive. Water is needed:

- for many chemical reactions inside plant cells,
- to keep the cells stiff (gives support),
- to cool a plant in warm conditions.

Water flow

Water is absorbed by the roots of plants. It moves up through a plant inside hollow tubes called xylem vessels (see Topic 2.1). Water loss from the leaves means that there is a lower water pressure here than in the root xylem. Since liquids always flow from points of high pressure to points of low pressure, water therefore travels up the xylem from roots to leaves. It then escapes through holes in the leaf surface. This flow of water is called the **transpiration stream**. The flow continues as long as water is being lost from the leaves. The faster this loss, the faster will be the transpiration stream.

Watering hanging baskets during the summer is a full-time job. These plants are watered twice daily from June to September, and three times on very hot days

Not enough water

The right amount of water is important. Plants quickly lose their support when they lack water. This is because their cells are no longer filled with the water that makes them turgid; the softer parts flop over. This is called wilting. The plants will die if they go much longer without water.

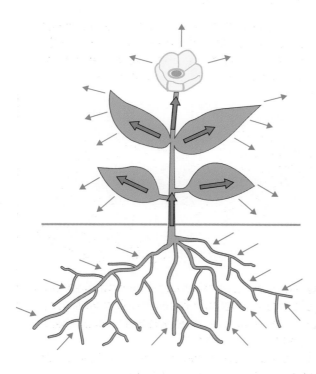

The transpiration stream

This plant is wilting from too little water

PLANT LIFE

Too much water

Most house plants die because they are overwatered! If they stand in too much water the roots suffocate. This is because water removes air from soil, taking away the oxygen needed for respiration (see Topic 2.8). Warm, very damp conditions may also encourage the growth of fungi, which can rot plants.

This plant has fungal rot from overwatering

★ THINGS TO DO

1 This picture, from a survival book, suggests how to collect drinking water.

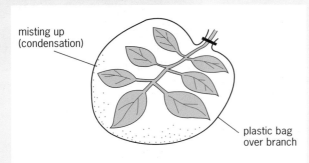

a) Why does condensation form in the bag?
b) Think of and write down some other ways of collecting water for survival.

2 Some plants are adapted so that they lose as little water as possible.

Rubber plants have thick, waxy leaves; pine trees have very small, waxy leaves; succulents have short stems and thick, waxy leaves

a) Explain how each feature helps retain water.
b) How could you study the effects of (i) leaf area and (ii) amount of wax on water loss?

3 Water may be stored in the root, stem or leaves for later use in dry weather. Succulents (as shown in question 2) store water in their leaves.
a) Succulents are adapted to growing in deserts. How do their leaves help?
b) Think of a plant which seems to store water.
i) How could you dry different parts of it to find out where most water is stored?
ii) How would you make sure that you dried out the parts completely?
iii) Write a report of your findings; compare it with others' reports.

4 The photograph shows special water-retaining chemicals that are added to soil to reduce the frequency of watering. They are helpful when people go away on holiday, but can be expensive.

a) Write a plan describing how you could carry out tests to find out which chemical was best.
b) If possible try out your ideas; display your findings on a poster.
c) How else could people prevent their plants dying while they were on holiday?

2.10 Supporting plants

Tall plants must be supported in some way. Young plants and many annual plants rely on water inside their cells and transport tissues to keep them upright. This means that cells which have filled with water by osmosis (see Topic 2.13) stay stiff (turgid), supporting the plant. Water travelling through xylem vessels and in phloem tubes also gives support. If cells lose water, the lack of support causes the plant to flop over and wilt.

The photomicrographs show the effects of water on cells.

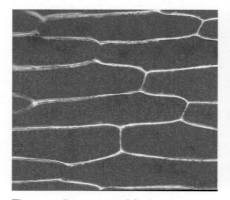

These cells are **turgid**; they have enough water inside to resist being squashed, and can support the plant

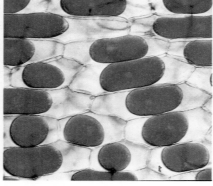

These cells are **flaccid**; they have lost water because the plant is drying out. Because there is less water inside the cell it is easily squashed by the pressure of those around it

These pea plants grow over 1 metre high. Their stems don't harden, so they depend on the water inside to stay upright. But this isn't enough when they've got the extra weight of the pods to carry and the wind blows. To give extra support, they can be grown around the netting

Better support

The stems and stalks of many plants become hard and much more supportive after a while. This happens because xylem vessels become filled with a very hard chemical called **lignin**. The hardened xylem is called **wood**. Water can no longer travel through this hardened xylem, so the job of the xylem cells has changed to one of support only. The diagrams show how this happens.

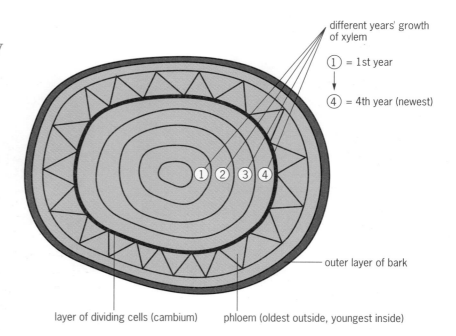

A stem section, showing wood formation

PLANT LIFE

New xylem vessels and phloem tubes are formed by cells in the **cambium** dividing. As new xylem is formed, new phloem tubes are also formed. Cells that are pushed inwards become xylem vessels, and those pushed outwards become phloem tubes. As the original phloem tissue is pushed outwards, the new phloem takes over the job of transporting food and hormones. As this happens the stem widens. The process is called **secondary thickening**. It also happens in the roots.

Secondary thickening in a stem

In deciduous trees, which lose their leaves in the autumn, new xylem and phloem grow every year. The evidence for this can be seen in the annual rings (see Topic 2.3).

★ THINGS TO DO

1 Cut leafy stems of plants like privet; immediately place them in water. Add a few drops of red ink or dye to colour the water. Leave for 1–3 days. The water travels up the xylem and colours it.

a) Carefully cut a section across the stem and use a hand lens to see where the colour is. Draw a diagram and label the xylem you identify.

b) Compare what you see with the stem sections opposite. Have you correctly identified the xylem?

c) This time cut a lengthways (longitudinal) section. Identify the coloured xylem. Draw a labelled diagram of what you see.

2 Test the strength of various twigs by clamping one end of them and placing masses on the other, as shown in the diagram. (You will need to assess the safety risk of using your equipment.)

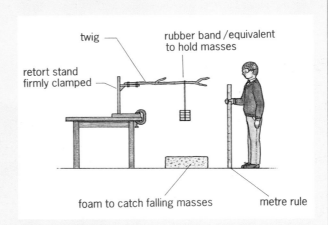

a) What might affect the strength of a twig?
b) Write down a hypothesis to test; include in it your reasons for what you think might happen.
c) Write a detailed plan of how to test your ideas, and have it checked.
d) Carry out your investigation, recording your results in a table.
e) Report your findings as fully as possible.

2.11 Water uptake

Root systems

As plants lose water from the leaves they replace it with water taken up through the roots. Growing plants need to absorb lots of water, so need a good root system. Different plants have different types of root system and are adapted to finding water in different ways, as shown in the diagram.

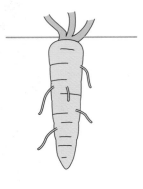

Carrots have long tap roots, with few side (lateral) roots, which can penetrate the deeper layers of the soil for water

Transplanting trees

Deciduous trees are best transplanted (moved from one place to another) in the winter, when they won't lose water so quickly because they have no leaves. Although the digging will destroy some roots, new ones will grow over winter, before the plant needs large amounts of water for growth in the spring. Steps must be taken to minimise damage to the root system.

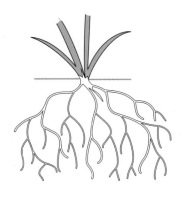

Grasses have a broad mass of fibrous roots that absorb rain water from a large area

Transplanting a deciduous tree

1 First dig around the tree, leaving plenty of room so few roots are damaged

2 Carefully lever out the large rootball

3 Dig out a hole big enough to take the rootball

4 Gently lower the tree in place, making sure it is at the same level as before

5 Add soil and firm into place. Water well as the new roots grow

PLANT LIFE

Root cells

Cells in a root are designed so that water absorbed from the soil passes into the xylem tissue in the middle. Root hair cells on the outside of a root have long 'arms', which grow between soil particles to find water. They also have thin cell walls so water can more easily enter. Their delicate structure means that they are easily destroyed, however, so new cells must continuously form as the root grows.

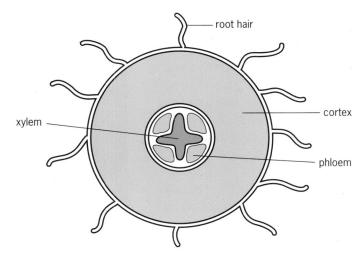

A root cross-section

Osmosis

Water enters into cells by a process called **osmosis** (see Topic 2.13). It is a slow process, so plants develop many roots and root hair cells to maximise their surface area. This is to enable them to absorb as much water as possible; it is a passive process (i.e. requires no energy). The diagram shows how water travels through the root.

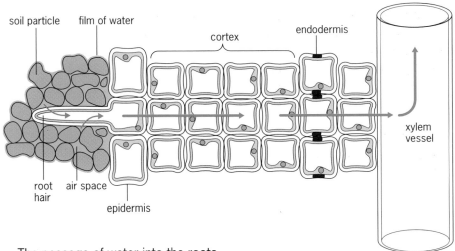

The passage of water into the roots

★ THINGS TO DO

1 Look at some plants collected from different areas.
   a) Draw the different types of root system. Describe the similarities and differences between each. For each one, add a comment saying how well you think it will help: (i) anchor the plant in the ground, and (ii) absorb the water the plant needs.
   b) Try to measure and estimate the total root length of each type of plant.

2 Germinate some seeds (e.g. mustard or cress). Use a hand lens to look closely at the root system that grows. You should be able to see the tiny root hairs. If possible, examine them more closely with a microscope. Draw a series of pictures showing how the root develops.

3 Trees growing in very cold places can suffer from a condition called 'physiological drought' when the ground freezes. Explain what you think this term means.

2.12 Losing water

Leaves lose water from the **stomata** (see Topic 2.5). These open during the day to allow carbon dioxide to enter the leaf and oxygen to escape. At the same time, water vapour is also lost. At night they close, so water cannot escape.

The opening and closing of stomata are controlled by **guard cells**. During the day these take in water and swell into banana shapes, causing them to separate and create a hole. At night they lose water and shrink, closing the hole again.

A stoma in a leaf, with guard cells

Measuring water loss

The amount of water loss from a leafy stem can be measured using a device called a potometer. As water escapes from the leaves more water passes up the stem to replace it. This draws water along the capillary tube, as shown in the diagram. By timing how quickly it moves along the tube, the rate of water loss can be measured (assuming that the rate of water flow through the potometer equals the rate of loss from the leaves).

The graph shows the results of an experiment to compare the amount of water lost in light and in darkness. They show that the water loss is greater in light.

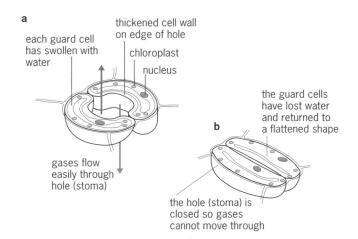

How stomata open and close: **a** daytime; **b** night-time

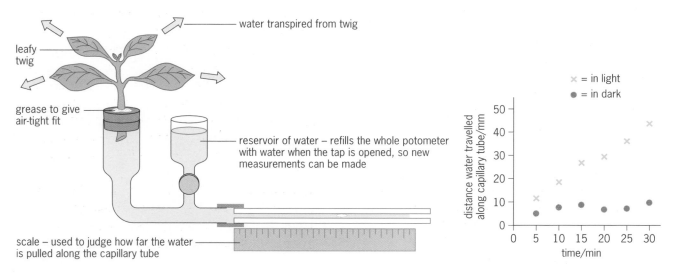

How a potometer works

Results of a potometer experiment

PLANT LIFE

Reducing water loss

Newly planted trees are often protected with plastic covers until they become established; this protects them from the drying effects of the wind

Plants lose water more quickly in warm, dry, windy conditions. If they lose water too quickly the leaves wilt, closing the stomata and halting further water movement until more can be absorbed by the roots. Plants survive by balancing the amounts of water that they take in and lose. Normally just enough is lost to keep water moving through the plant. Most plants have more stomata on the sheltered underside of their leaves.

sunlight reflects off this leaf because the surface cells have a thick covering of waxy cuticle – this makes it harder for water to pass through

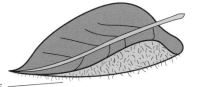

hairs trap air, so the humidity increases; less water is lost in highly humid conditions

the leaf of this marram grass plant has curled up – the air trapped inside becomes very humid, so less water is lost

Plants growing in the drying conditions of a shingle beach are adapted to reduce transpiration as far as possible. They have a number of special features

---

★ THINGS TO DO

1 The impression of stomata can be seen under a microscope if the surface of a leaf (e.g. a privet) is painted with a thin layer of clear nail varnish. When dry the varnish can be pulled off the leaf and mounted on a drop of water on a microscope slide.
   a) Prepare slides from each surface of a leaf; compare them under a microscope.
   b) Make a table of the similarities and differences between the two surfaces.

2 Nail varnish can be used to block the stomata so water cannot escape. If painted leaves are left in dry conditions they will lose water only from unpainted areas. The leaves lose weight and may change shape as they dry out.
   a) How could you use this technique to decide which surface loses most water?
   b) Write a detailed account of what you did; explain what you found out.

3 The graph opposite shows that more water is lost in light than in darkness.
   a) How can the results be explained using the idea of stomata on the leaves?
   b) Jess thought that readings for the dark were inaccurate. What might have caused errors?
   c) How could she obtain more reliable results?
   d) How could the potometer be used to test the effect of other factors on the transpiration rate?
   e) If possible, investigate your ideas. You might first try using data-logging equipment to measure how factors such as moisture and light levels vary over 24 hours, or to measure water loss.

2.13 On the move

Chemicals move into and out of plant cells in three different ways:
(1) osmosis,
(2) diffusion,
(3) active transport.

Osmosis

The movement of water by osmosis was introduced in Topic 2.11. It is important to remember that this process does not need energy from the plant. It is known as a 'passive process'. In other words, it does not need the energy released during respiration. Osmosis involves the movement of water molecules from one solution of chemicals to another, separated from the first by a barrier that allows only the water molecules through. Water molecules will always move from a less concentrated into a more concentrated solution to balance the numbers of water molecules on each side. They continue to do this until the two solutions are at the same concentration (or until pressure stops further movement).

Water can enter or leave cells by osmosis because the cell membrane acts as a barrier to larger molecules (a partially permeable membrane). Water molecules enter a root hair cell because the concentration of the solution inside the cell is greater than that outside; they move along a 'concentration gradient'. As they enter they lower the concentration of the solution in that cell. Because of this, water molecules then move into the next cells, where the concentration is still at the original strength. In this way, water passes from cell to cell inward towards the xylem tissue in the middle of the root. Stomata in the leaves open and close when osmosis causes the guard cells surrounding them to change shape.

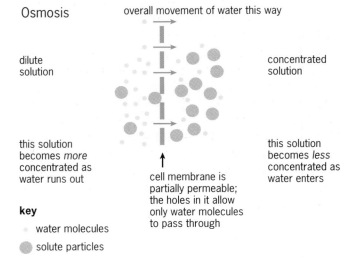

Osmosis

Diffusion

Diffusion is also a passive process. It was explained in Topic 1.10. Minerals in solution diffuse into root hair cells because their concentration in soil water is higher than that in the cell sap. Once absorbed into the root hair cell, they move further into the root by diffusion.

Gases move more easily by diffusion than do liquids. Water vapour diffuses out of leaves during transpiration. Carbon dioxide and oxygen also diffuse into and out of leaves. The direction in which they move depends on their relative concentrations inside and outside the leaf. The diagram shows their movements by day.

Particles moving from high to low concentration into a root hair cell by diffusion

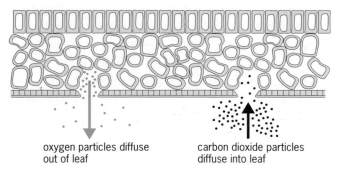

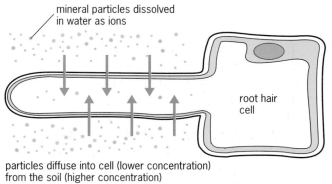

O_2 and CO_2 movements during the day

PLANT LIFE

Active transport

Plant roots also absorb minerals partly by a process which requires energy from the plant itself, and hence is called active transport. This is needed when the difference in concentrations would tend to make them move the opposite way by diffusion. In other words, active transport enables the minerals to move *against* a concentration gradient, from a low concentration to a high concentration, as shown in the diagram. The energy needed is released from food by respiration so the cells involved in active transport have lots of mitochondria (where respiration occurs – see Topic 1.2); these provide the extra energy needed.

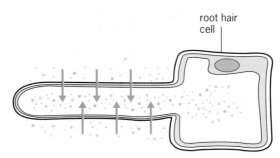

particles are forced into the cell by a type of 'pumping' mechanism – this needs energy

Active transport: intake of mineral ions against a concentration gradient

★ THINGS TO DO

1 a) Copy the diagram of a leaf showing gases diffusing during the day.
b) At night the leaf loses carbon dioxide and takes in oxygen. Draw another diagram to explain how this happens.

2 Look at the bar chart showing the relative amounts of various minerals in pond water and in the cell sap of a microscopic green alga.

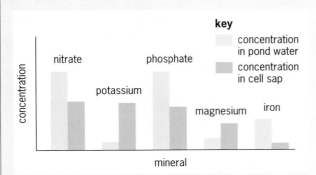

a) Which minerals can be absorbed by diffusion? Explain your answer.
b) Which minerals must be absorbed by active transport? Explain your answer.
c) When soil warms up in the summer, minerals move into the roots more easily.
  i) Why is diffusion quicker in warm conditions?
  ii) Why is active transport quicker in warm conditions?

d) As a plant absorbs minerals from soil the concentration gradient changes. Why does a plant need more energy to absorb minerals as they are depleted in the soil?

3 Jenny works in a burger bar, making French fries. She keeps the cut potatoes in water to stop them going brown before being cooked. She noticed that they went soft if left in salty water, but went hard again if moved into fresh water.
a) Potato chips consist of potato cells. Water inside the cells keeps them stiff. Why do the chips soften in salty water?
b) Why do they reharden in fresh water?
c) How could you test your ideas using salty water at different concentrations?
d) If possible, try this out. Explain your findings in terms of water movement by osmosis.

4 Diffusion and active transport are important processes in animals, including humans, as well as plants. (Look back through Topics 1.7, 1.10 and 1.18 to answer these questions.)
a) Which gases move by diffusion in the lungs?
b) Explain how sugar and digested food molecules diffuse into blood in the small intestine.
c) Which chemicals are moved by active transport in the kidneys?

2.14 Controlled growth

Plants that are not subjected to human influences like this bonsai tree can still grow into curious shapes.

Natural control

The growth and shape of plants may be affected by many different factors. The three most important environmental factors are:

- light,
- water,
- gravity.

Plants respond to changes in each of these factors by growing in a certain direction. The response is called a **tropism**.

Light

The first plant in the illustration (right) is responding to light reaching it from one direction only. Its stem is growing towards the light. This allows the leaves to receive a maximum amount of light for photosynthesis and makes the flowers very obvious to pollinators such as bees and insects.

The response to light is called **phototropism**. Stems grow towards light, so are said to be positively phototropic. Roots are often unaffected by light, but if they do respond they grow away from it. So they are negatively phototropic.

The response of stems and roots is controlled by a chemical. This chemical is a plant hormone called **auxin**, which affects growth. Different concentrations of the hormone across the stem stimulate it to grow at different rates on opposite sides of the stem, causing it to bend, as the diagram opposite shows.

Water and gravity

Roots tend to grow towards water (see illustration above). This response is called **hydrotropism**. Roots are said to be positively hydrotropic.

Bonsai trees are miniature versions of trees that usually grow to a height of 30 metres. Their growth has been stunted by keeping them in shallow pots and pruning the roots regularly; also new branches are carefully pruned to a desired shape (some people also twist wire around the trunk and branches to achieve this)

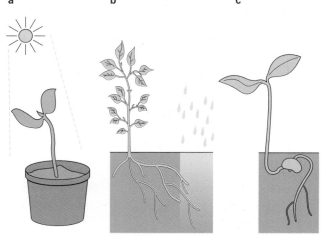

Directions of plant growth: **a** young plants bending towards a unidirectional light; **b** plant roots searching out water; **c** stem and root bending away from and towards the ground respectively

Shoots grow upwards from a germinating seed. The roots, however, always grow downwards. The plant responds to the gravitational pull of the earth; this response is called **geotropism**. Roots grow in the direction of gravity (positive geotropism) and shoots grow against gravity (negative geotropism), as shown in the illustration above.

PLANT LIFE

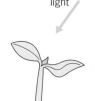

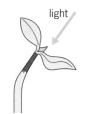

1 Direction of light detected at top of stem

2 More growth hormone on side of stem furthest from light

3 More growth on shaded side of stem results in stem bending toward light

4 Growth hormone now passes evenly down stem – so stem continues to grow towards light

How a stem bends towards the light

★ THINGS TO DO

1 a) Explain why a plant is more likely to survive if it grows towards the light.
b) Why do plants in the open grow straight upwards and not to one side?
c) How could you test your ideas by using a young plant or seedlings?
d) Seedlings sense where light is coming from and quickly respond to any change in direction. You could investigate this using foil to shade parts of seedlings from light.
i) Which part of a seedling is most likely to sense light? Why?
ii) Write a plan to test out your ideas; have it checked and try it.
iii) Report on your findings; explain their significance.

2 Experiments with oat seedlings, as seen in the diagram, have shown that a chemical made in the shoot tip stimulates growth. The chemical travels down the tip by diffusion, causing cells just below the tip to grow. Increasing the amount of the chemical increases the growth rate.
a) Explain why the shoot stops growing when the tip is cut off and removed.
b) Why does the shoot carry on growing if the tip is replaced?
c) Why does the shoot bend when the tip is replaced on one side?
d) Try to use these results to explain why a chemical hormone is thought to control the growth of a shoot towards light coming from one direction.

e) How could you use a sample of auxin to check whether it can alter growth in oat seedlings?

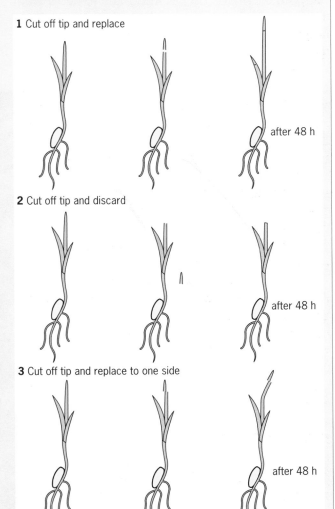

1 Cut off tip and replace

after 48 h

2 Cut off tip and discard

after 48 h

3 Cut off tip and replace to one side

after 48 h

2.15 Changing nature's way

Humans can use the chemicals (including hormones) that control what plants do to stay alive and healthy in order to produce 'perfect' plants. Artificial chemicals are used to control what plants do: how they grow, and when, to suit our convenience. They can be used for:

- stimulating the growth of roots when plants are being propagated from stem cuttings (see Topic 4.7),
- slowing down or speeding up the development of fruit for human consumption,
- disorganising growth in weeds and so killing them.

Using chemicals like this can have hidden dangers, however, so it is important to know what the potential problems are. Regulations ban the use of certain chemicals that are known to have health and environmental problems.

Slow boat of bananas

Bananas are good for you, but they're not as natural as you might think! Unripe green bananas are picked in tropical countries such as Belize. To prevent them ripening during the 3-week boat journey, they are stored at fairly low temperatures (13 °C). Once in Britain, ripening is speeded up by raising the temperature of their storage room to 22 °C and adding a gas called ethylene (ethene) to the atmosphere.

Ethylene, C_2H_4, is normally produced inside the banana and switches on the natural processes of ripening. It does this by stimulating the production of enzymes which cause:

- the skin colour to change from green to yellow (and later black),
- the development of chemicals that give a banana its flavour,
- the texture of the banana to soften.

Green and yellow bananas

The right combination of ethylene and temperature varies the ripening time, but once on the shelves bananas deteriorate after a week. Sealing in plastic bags modifies their surrounding atmosphere and can delay over-ripening. Storage life is extended if carbon dioxide builds up, inhibiting respiration and making enzymes work less effectively. Adding potassium permanganate to destroy any ethylene made naturally by the bananas keeps them fit to eat for even longer.

Rooting powder

Rooting powder contains an artificial form of the hormone auxin. This stimulates the growth of roots from the cut stems of cuttings. It adds to the effect of a natural hormone that accumulates at the base of leaf stalks, so quick growth is more likely.

Hormones are also used to produce new plants by tissue culture (see Topic 4.8).

PLANT LIFE

Weedkillers

Selective weedkillers are useful to kill plants like clover, dock and dandelion growing in lawns or on fields of cereal crops. Their big advantage is that they do not kill the grass or the crops – only the weeds. This is because these weeds have broad leaves, which absorb more of the weedkiller as it is sprinkled on. Grasses, including cereal crop plants, have narrower, more erect leaves so the weedkiller easily runs off.

Effect of a selective weedkiller on broad-leaved plants

Selective weedkillers contain synthetic compounds, made in chemical factories. Some compounds mimic the effect of natural growth-stimulating plant hormones (e.g. auxin). The weeds receive a massive overdose of growth-stimulating chemicals, and normal cell activity is disrupted as the growth pattern is changed. The plants grow themselves to death!

Artificial daylight

Using artificial lights to extend the daylength is one way to trick plants into growing or flowering at certain times, for instance the production of daffodils or other flowering plants for Christmas. The response is controlled by hormones.

★ THINGS TO DO

1

The photograph shows a bare patch that was good lawn before excess weedkiller was spilled on to it.
a) Why has the grass died in this part of the lawn?
b) Find out what dose of weedkiller is recommended for use on a lawn.
c) How can this amount be spread evenly on a lawn?
d) Why is that important?
e) What safety precautions are needed by people applying weedkillers?
f) How could you investigate the effect of using different doses of weedkiller on weeds? Write a plan for this long-term test; carry it out if possible.

2 Cut stems from plants like geranium will develop roots if left in a container of water for several weeks. Some may die or go mouldy in this time.
a) What causes the stem to develop roots?
b) What are the advantages of using hormone rooting powder?
c) What should you check before using this powder?
d) i) List the factors that could affect the growth of new roots.
ii) How could you test your ideas?

Exam questions

1 a) The diagram shows a geranium plant.

i) Use the words in the list to label the diagram.

flower   leaf   root   stem (3)

Plants produce food by the process called photosynthesis.
ii) Where does photosynthesis take place in a plant? (1)

iii) Write down **two** things plants need for photosynthesis. (2)
iv) Name the process by which plants release energy from food. (1)

(MEG, 1995 (part))

2 a) The diagram shows two young plants grown by Blodwen. One was grown on a window sill, the other in a dark airing cupboard.

i) Which plant was grown in an airing cupboard?
ii) Give **one** reason for your answer. (2)
b) i) She left plant **B** undisturbed near a window for several weeks. [In the space left] Draw the appearance of the plant after this time.

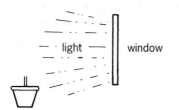

ii) Why is this reaction beneficial to the plant? (2)
c) Suggest **one** advantage of reproducing plants by cuttings over sexual reproduction. (1)

(WJEC, 1992 (part))

3

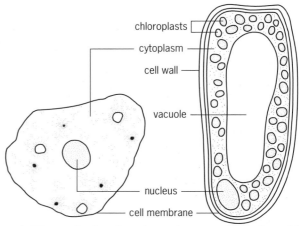

a) Use **only** the drawings above to fill in the table below to show **three** differences between plant and animal cells (3)

Plant Cells	Animal Cells

b) What is the job of **each** of the following parts of a cell: (2)
i) nucleus;
ii) cellulose cell wall;
iii) cytoplasm;
iv) cell membrane?
c) i) What substance is contained in the chloroplasts? (½)
ii) Why is this substance important to **all** living things? (½)

(WJEC, 1994 (part))

4 Plants lose water through their leaves.
a) i) What name is given to the loss of water from the leaves? (1)
ii) What name is given to the pores through which this water is lost? (1)
iii) Explain why the movement of water through plants is important to them. (3)
b) The loss of water from a leafy shoot can be shown using a potometer. A potometer is shown in the diagram below. As water is lost from the leaf the bubble slowly moves along the scale from left to right.

What will be the effect on the movement of the bubble of:
i) increasing the temperature of the air around the leafy shoot; (1)
ii) increasing the humidity of the air around the leafy shoot; (1)

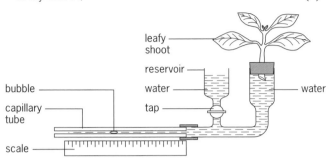

iii) opening the reservoir tap? (1)
c) Plants which live in dry, desert-like areas often have leaves which are modified to form sharp spines or prickles. State **two** ways that these modified leaves help the plant to survive in the desert. (2)
(SEG, 1995)

5 The drawing shows a sectional view of a root hair on a root in some soil.
a) One of the jobs of a root hair is to take in water from the soil.

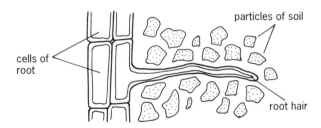

State **one** feature of the structure of the root hair. Explain how it helps the root hair to take in water. (2)

b) Describe, in detail, how water passes from the soil into the root hair. (4)
(SEG, 1994)

6 The diagram below represents the flow of chemicals and reactions taking place in the leaf.

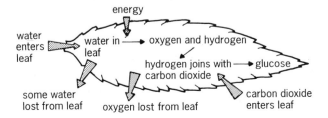

a) i) Where does the energy come from? (1)
ii) What **name** is given to the process of water loss from leaves? (1)
iii) Carbon atoms with a mass number of 14 are radioactive. Describe how carbon-14 could be used to prove that glucose in a leaf has been made from carbon dioxide. (3)
iv) In which part of the leaf do the reactions shown in the diagram take place? (2)
b) State **three** ways the plant uses glucose which is made in the leaf. (3)
(SEG, 1994)

7 The figures below show how the yield of a wheat crop is affected by adding nitrogen fertiliser.

Nitrogen fertiliser added (kg/hectare)	Yield of wheat (tonnes/hectare)
0	26
50	28
75	31
100	34
125	40
150	43
175	44
200	44

a) Display these results on graph paper in the most suitable way. (4)
b) How much nitrogen fertiliser would need to be added to the wheat crop to get a yield of 37 tonnes/hectare? (1)
(NEAB, 1994 (part))

8 The diagram below shows a section through a leaf.

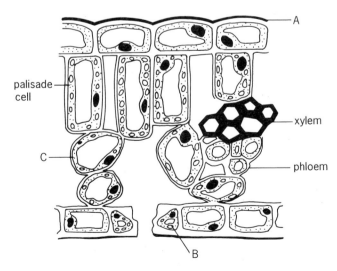

a) Label layer A and cells B and C [on the lines provided]. (3)

EXAM QUESTIONS

b) Describe the function of the chloroplasts in the palisade cell. (2)
c) What is the function of the xylem in a leaf? (1)
d) What is the function of the phloem in a leaf? (1)
(ULEAC, 1995)

9 Some students are studying a plant which climbs on both sides of a wall in their school grounds.
One side of the wall faces north and the other faces south.
The students measured the length and the width of 100 leaves from each side of the wall.
They used these measurements to calculate the average length and width of the two groups of leaves.
The table shows their results.

	south facing (sunny) side	north facing (shady) side
average length of leaf/mm	63	102
average width of leaf/mm	41	83

a) The rate of carbon dioxide exchange by some of the leaves was measured at different light intensities.
The graph shows the results.

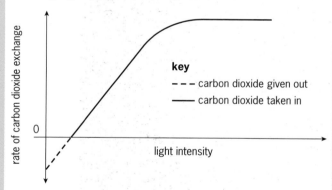

Describe how the rates of photosynthesis and respiration would vary on the sunny side of the wall from midnight to midday on a cloudless day. (5)
b) The students observed that when the shoots of the plant on the shady side of the wall reached the top of the wall they all grew towards the light.
This response is called positive phototropism.
Explain the value of phototropism to plants. (2)
(MEG, 1995)

10 The diagram below shows tomatoes being grown in a glasshouse.
Group A were grown in a 'grow bag' containing compost. Group B were grown in a tube through which organic 'slurry' flowed constantly.

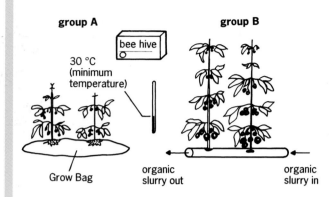

The tomato plants in Group B were much taller and more productive that those in Group A.
a) Explain how the factors shown in the diagram affect the growth and productivity of both groups of tomato plants. (4)
b) Even though tomato plants can flower and fruit at any time of the year, it has proved uneconomic to grow tomatoes in glasshouses in Britain between November and February.
Suggest **two** reasons why this is so. (2)
(ULEAC, 1994)

11 A scientist measured the percentage of carbon dioxide in the air one metre above the centre of a wheat field. It was measured on a calm day every three hours during a 24-hour period. The results are shown on the graph.

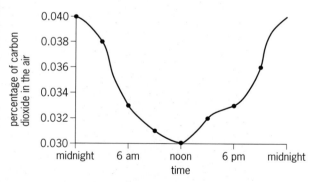

a) What is the change in the percentage of carbon dioxide in the air between midnight and 6 a.m.? (1)
b) The amount of carbon dioxide in the air above the wheat field changes. Why does it change in the way shown by the graph? (2)
c) The amount of oxygen in the air one metre above the wheat field changes. How does it change during this same 24-hour period? Explain your answer. (2)
d) Green plants can store starch. Describe a simple test which shows the presence of starch. (2)
(SEG, 1995)

THE ENVIRONMENT

3.1 It's our environment

All around the world, living things exist alongside one another. There are many different types of place – called **habitats** – in which plants and animals live. These provide everything that living things need to survive.

Dry, wet, hot and cold habitats

Climate

Climate affects all living things. Plants need water, nutrients, air and warmth to grow quickly. Animals need air, food, water and warmth for their survival. Some can live in dry areas; others can only survive in areas that are permanently wet. In the warm, damp conditions of the tropical rain forest, plants grow quickly, providing an abundance of food for animals. So these habitats contain a much wider variety of living things than the freezing cold of Antarctica. Some people think that there are plants and animals living in rain forests – well away from human influences – that have never been discovered. Some of those plants may contain drugs that could help cure many diseases that are currently fatal to humans.

Ecosystems

As plants grow they absorb nutrients from the soil. The nutrients are replaced as microbes in the soil break down animal wastes, and dead plants and animals. Other living things (earthworms, for example) improve the quality of the soil – burrowing through it, and dragging leaves and other plant material into it. Growing plants' roots also hold the soil's particles together so it is not eroded and blown away by the wind.

Humans, and other animals, rely on plants for food. Plants provide one other thing that we all need – oxygen. Without them the atmosphere would not be able to support life as we know it. They also serve a second purpose: removing carbon dioxide from the air and turning it into food.

All living things are part of an ever-changing system – an **ecosystem**. Within an ecosystem plants and animals grow, reproduce and die. Within any habitat, plants and animals interact. Animals eat plants and may themselves be eaten by other animals. Animals may move to a new area that has better conditions, or move out if, for example, the food supply cannot support their needs. Plants may die out owing to climate changes, lack of nutrients, pollution or because they have been eaten. Others may move in, followed by the animals that feed on them.

Human influences on the environment

Humans have an effect on the surroundings (the environment) in all sorts of ways. Like other animals and plants, we need certain things to stay alive. We build homes and factories, quarry the earth for rocks and drill for oil and gas. Land is farmed for food, and the sea and rivers are fished. We release wastes into the air and water and on to the land, causing pollution. Forests may be chopped down or planted.

THE ENVIRONMENT

Some of the things we do improve the environment; others do harm. Some plants and animals may find the new conditions make it easier for them to survive. Others may not. Sadly, the survival of many types (species) of plants and animals is threatened by what we do. Human influences can be good (positive) or bad (negative) and can have long-term or even permanent effects, including the extinction of species.

Human influences on the environment: **a** animal farming; **b** quarrying and land refill; **c** power stations; **d** waste disposal; **e** forestry; **f** fuels

★ THINGS TO DO

1 What is meant by the following terms:
  a) habitat;
  b) ecosystem;
  c) environment;
  d) human influences;
  e) species;
  f) extinction?

2 The charts show the typical climate in southern Britain over a year (red = temperature).

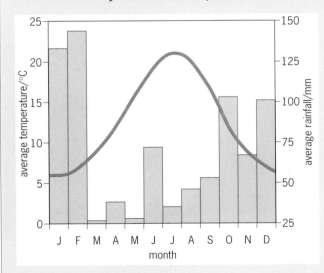

  a) Which season is the warmest?
  b) Which season is the wettest?
  c) How do you think changes in the seasons affect the animals and plants that live there?
  d) Some people say that the climate is changing owing to 'global warming'. If the British climate became warmer how would that affect the environment?

3 The photographs above show some ways in which humans influence the environment.
  Think of the possible effects of each human activity on plants and animals. List the effects as good or bad. For example, on farmland:
  good effects – good conditions for keeping animals, so some wild animals can survive; good soil conditions for growing plants, so some wild ones also grow;
  bad effects – wild plants are weeded out; many wild animals like rabbits and foxes are shot; some fields are drained, so wild plants and animals that like these conditions cannot survive.

3.2 Places to live

Britain has a wide variety of habitats, partly because it is an island and also because the different rocks from which our island is made produce different types of soils. Some habitats are natural, formed by the forces of nature and the influence of living things. Others, such as reservoirs, are artificial, made by humans.

All of these habitats offer different conditions for plants and animals, and so contain different mixes (communities) of living things.

Soil

Most plants grow best in rich soil. Soil is formed as rock is broken into tiny pieces by the weather (wind, rain, cold etc.). Different types of rock produce different soil types. Some plants prefer to grow in soil made from weathered limestone rock, which is alkaline. Plants growing on a limestone pavement, as shown in the illustration, are in a very special habitat. Many plants found there will grow nowhere else. Other plants, such as heathers and rhododendrons, prefer soil that is acidic (often found on moors).

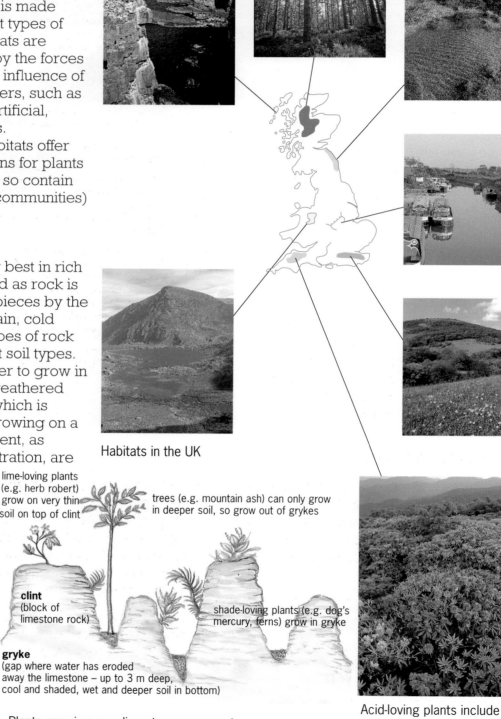

Habitats in the UK

lime-loving plants (e.g. herb robert) grow on very thin soil on top of clint

trees (e.g. mountain ash) can only grow in deeper soil, so grow out of grykes

clint (block of limestone rock)

shade-loving plants (e.g. dog's mercury, ferns) grow in gryke

gryke (gap where water has eroded away the limestone – up to 3 m deep, cool and shaded, wet and deeper soil in bottom)

Plants growing on a limestone pavement

Acid-loving plants include rhododendrons

THE ENVIRONMENT

At camp

Simple surveys can show the variety of plants and animals that can be found in any habitat. The diary on the right was written by a student who went camping in May. Part of his task was to keep a note of the conditions, and of the animals and plants that live there.

Wednesday

4 p.m Arrived and set up camp. Wanted to pitch the tent near the pond, but too wet and mossy down there. Chose a site in short grass, with shade from some willow trees.

7 p.m. Went to look at the newts in the pond. There were lots. Plenty of midges too - used cream to stop them biting. A heron flew in and speared a frog from the edge of the pond!

11 p.m. Lots of noise from animals calling in the dark. John is a birdwatcher and says that the whirring we can hear is a nightjar - a bird that has migrated from Africa to breed here.

Thursday

2 a.m. Finally got to sleep, then woken by hooting coming from the wood - a tawny owl. Went to the toilet - lots of moths flying around the lights in the toilet block.

5 a.m. Woken by dawn chorus. Why have the birds got to sing so early?

8 a.m. Had breakfast sitting on an old log outside the tent until a woodlouse crawled up my leg - ugh! Found stacks of them under the log, with lots of beetles, slugs and other bugs. I'll not sit there again in a hurry. Went for a wash - moths had gone.

10 a.m. Went for a walk in the wood, but no sign of the tawny owl. It is quite dark and cold under the trees. Apart from the trees there aren't many plants. There are a few ferns and foxgloves. John says that some of the singing birds are migrants. They are singing to attract a mate so they can breed. He's spotted a lot of willow warblers and a flycatcher. I saw two squirrels. Lots of caterpillars on the tree leaves.

12 p.m. Back at the pond, but saw no newts. Some tadpoles swimming in the water, eating the pondweed. Butterflies and bees flying over the pond and pondskaters walking on top of the water.

★ THINGS TO DO

1 Copy the table below into your notebook and name animals and plants that live in each place. Use the diary above as a source of information, but try to add some plants and animals that you have seen in each habitat. If possible, have a look for yourself.

Habitat	Plants	Animals
pond		
wood		
grassed area		
rotting log		

2 Walk around the school grounds or a local area such as a park.
a) List the different habitats you see, and the plants and animals that live there.
b) Make a list of the conditions needed by plants and animals, and describe how each habitat provides them.

3 Draw a map of your local area and locate as many different habitats as you can. Include both natural and artificial habitats. You might visit later to study what lives there and why.

3.3 Matching the conditions

Plants and animals tend to stay in one type of habitat because the conditions there increase their chances of survival. They are usually adapted to living in a particular habitat.

Some trees have broad leaves to absorb as much light as possible

Conifer trees have very small leaves to reduce the amount of water they lose

Owls have special wing feathers that allow them to swoop silently down for prey, sharp talons to grasp their prey, and sharp, hooked beaks to tear meat

Tadpoles have gills to absorb oxygen from water as they breathe. As tadpoles develop into frogs their breathing system changes – they develop lungs

In the wood

Plants try to outgrow each other so that they get enough light for photosynthesis. Those living under trees in a wood have many problems. Trees have extensive root systems, which absorb most of the water and nutrients from soil, leaving little for other plants. They also put other plants in the shade. Since all green plants need light to make food, any living in the shade must have adapted in some way in order to survive in those places. In a wood with deciduous trees, smaller plants survive only because the trees lose their leaves for part of the year (see Topic 2.4). Different plants grow and flower at different times to take advantage of the available light, as shown in the illustrations and the table beneath it.

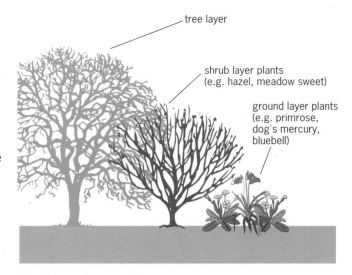

Plant layers in a wood

Bluebells grow up, do most of their photosynthesis and store most of next year's food before the buds open on the trees above. This maximises their use of the limited amount of light

Month	J F M A M J J A S O N D
Maximum shade under trees	_____
Time when plants grow and flower:	
hazel	_____
primrose	_____
dog's mercury	_____
bluebell	_____
meadowsweet	_____

Growing and flowering times for different woodland plants

Life in the saltmarsh

Animals and plants in a saltmarsh

A saltmarsh is an area of low ground that the sea covers at high tide, so the plants growing there must be able to survive the salty conditions caused by seawater. The diagram shows some of the variety of animals that can be found there. Snails and geese both feed on the eelgrass, and brine shrimps can be found in pools of salty water. A number of birds use saltmarshes as feeding grounds, including birds of prey such as harriers and peregrine falcons, which seek out other birds for food. Waders – birds with long legs – move through the water as they feed. At low tide, when the mud is exposed, they probe the mud with their bills to find food. Different waders are adapted to finding animals at different depths in the mud and water so have different bill sizes and shapes.

★ THINGS TO DO

1 a) Explain what 'being adapted' means and why it is important.
   b) What problems do plants have growing under trees?
   c) Why can plants grow underneath deciduous but not under coniferous trees?
   d) Which plant in the table can grow in the shade?
   e) Explain how dog's mercury manages to grow under trees.

2 Look at the waders in the diagram of the saltmarsh.
   a) Why are long legs useful to them?
   b) What special feature helps curlews to collect most food?
   c) Explain how each bird is adapted to feeding on different animals.
   d) Why do curlews have the best chance of finding animals to feed on?
   e) Curlews have good eyesight, and can see behind themselves without turning their heads. How will this adaptation help their survival?

3.4 Using keys

A biological key can be used to identify living things. Sometimes the key is a series of photographs or diagrams. More commonly, it is a list of questions or statements about the organisms to be identified. If you answer the questions in a systematic way you will eventually end up with the name of the organism. Try to follow these branching keys.

Key for animals found in fresh water

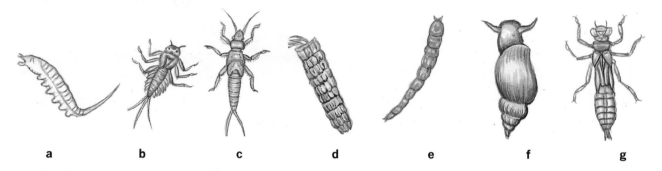

a   b   c   d   e   f   g

Some freshwater animals

1 tail has long extensions . . . . .go to 2
    no extensions on the tail .go to 4
2 tail has one extension . . . .**rat-tailed maggot**
    tail has more than one extension . . . . .go to 3
3 tail has 2 extensions . . . .**stonefly nymph**
    tail has 3 extensions . . . .**mayfly nymph**
4 animal is inside a case or shell . . . . .go to 5
    animal is not inside a case or shell . . . . .go to 6
5 case made of many small pieces . . . .**caddis fly larva**
    shell is made of 1 piece . . . .**water snail**
6 animal has 3 pairs of legs . . . .**dragonfly nymph**
    animal has no legs . . . .**bloodworm**

To use this key, look at one animal, then ask each question and answer in turn. For example, look at animal **a**:

1 Does the tail have long extensions on it? . . . . . . .Yes, so go to 2
2 Does the tail have only one extension? . . . . .Yes, so it is a rat-tailed maggot
and so on.

Now look at animal **b**:

1 Does the tail have long extensions on it? . . . . . . .Yes, so go to 2
2 Does the tail have only one extension? . . . . .No, it is not a rat-tailed maggot
   Does the tail have more than one extension? . . .Yes, so go to 3
3 Does the tail have two extensions? . . . . .No, it is not a stonefly nymph
   Does the tail have three extensions? . . . . .Yes, it is a mayfly nymph
and so on.

Key for plants growing in grassland

The following is a different type of key, in which you follow the diagram down from top to bottom.

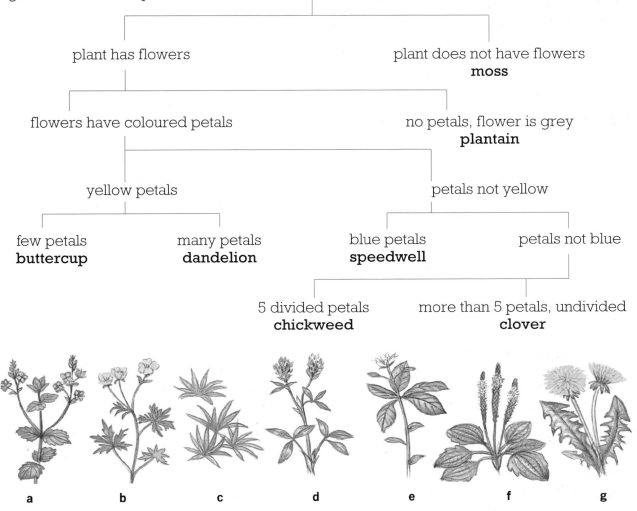

Some grassland plants

★ THINGS TO DO

1 Look at the key for water animals.
  a) Use the key to identify animals c to g in the diagram.
  b) Try to write another key for these animals, based on their size.
  c) If possible, take a sample from a stream (see Topic 3.5) and use the key to identify any animals you find.

2 Look at the key for plants in a lawn.
  a) Use the key to identify plants a to g in the diagram.
  b) Rewrite this key in the same form as that used for freshwater animals.
  c) Plants are not always in flower. Write a key for plants that would be useful in winter (the leaves would still be seen then).
  d) If possible, collect plants from a grassed area; display them with your key as a poster.

3 Find some friends with different pets. Write a key to identify these animals. Use photographs, drawings or descriptions as a starting point. Display your key as a poster.

3.5 Studying habitats

A square frame called a **quadrat** can be used to mark out an area to study the distribution of animals and plants. By identifying and counting animals and plants living in different parts of a habitat you can gain information about how their distribution varies. The method shown in the diagram is called a **belt transect**.

The results from a belt transect carried out above the shoreline of a beach are shown in the table.

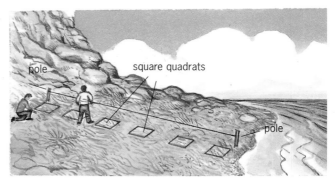

A belt transect on the seashore

Distance/m	60	50	40	30	20	10	0
barnacle					*	*	
winkle		*	*				
mussel				*	*		
sea slater	*						
sandhopper	*	*					
shrimp					*	*	
sea anemone					*	*	*
limpet		*	*	*	*	*	

You can see that different animals live at different distances from the shore. A belt transect from a hedge into a field of grass would show that different plant species grow as the amounts of light, water and shelter change.

Animal ways

Studying the distribution of most animals is difficult because they tend to move around. Some must be caught before you can identify and count them, as shown in the diagram.

Physical conditions such as temperature, light and dissolved oxygen can also be measured. Computer sensing devices can be used to measure changes in factors such as these over 24 hours or more, which might affect the plants and animals living there.

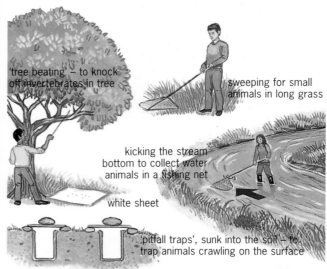

Ways of sampling different places for animals

Some instruments used to measure environmental conditions

THE ENVIRONMENT

Counting what you don't see

It is not always possible to catch and count every animal living in a habitat. The problem can be overcome by using a technique called **capture–mark–recapture**. Imagine you wanted to know how many woodlice live in a wet area with rotting wood. You would capture as many as you can, mark them with a small spot of enamel paint and release them back into the area. You return 1 day later and again capture as many as you can. Now you count separately those that were marked the previous day.

Birds like this black kite are ringed to follow their population changes

Suppose on the first day you captured and marked 156 woodlice; on the second day you captured 300, of which 35 have marks painted on. The mathematical equation shown below is used to calculate the number of woodlice in the area. It is called the **Lincoln index**. (It assumes that the number of marked woodlice caught are in the same proportion as the woodlice originally marked by paint.)

$$\frac{\text{number of marked woodlice caught}}{\text{total number of woodlice caught on 2nd day}} = \frac{\text{number of woodlice marked with paint}}{\text{total number of woodlice}}$$

Using this equation, the estimated number of woodlice in the area would be 1337.

★ THINGS TO DO

1 Different-sized quadrats were used to sample weeds in a grassed area.

Size of quadrat/m²	Number of different weed species found
0.2	3
0.5	6
0.8	8
1.0	11
1.5	11
2.0	10
4.0	11

a) Plot the data on a line graph.
b) Why is a 1 m² quadrat the best size to use to sample plants here?
c) How could quadrats and sensing devices be used to look for patterns in plant distribution depending on physical factors such as light, slope, amount of water in the soil or shelter from wind?
d) If possible, do a survey. Write a full report of what you find.
e) See whether the number of earthworms in soil matches the distribution of weeds. (Earthworms can be counted after soapy water is poured on to soil inside a quadrat. Wash the earthworms in clean water and hide them from predators like birds when you have counted them.)

2 Look at the results of the beach survey opposite.
a) Describe where different animals live on the beach.
b) What physical factors could affect where these animals live?
c) Green seaweeds grow at the top of the beach; brown seaweeds grow further down. How could you look for patterns in where seaweeds grow on the beach?

3.6 Living together

All living things need food. Food provides the energy and chemicals essential for life. Plants use energy and chemicals from their surroundings directly to make food. The energy comes from the Sun; the chemicals come from the air and the soil. Animals cannot do this; they must get their energy and chemicals from their food. Animals (including ourselves) ultimately depend on plants for survival.

We can link organisms together by thinking about the food they eat. The apple in the photograph came from an apple tree, which made its own food. The apple is food for people.

This connection can be shown as:

Food from **a** a plant; **b** an animal

$$\text{apple} \longrightarrow \text{person}$$
(food stored in the apple)   (supplies food for the person)

The milk was made from the grass that was eaten by the cow.

$$\text{grass} \longrightarrow \text{cow} \longrightarrow \text{person}$$
(food made by the plant)   (grass is food for the cow, which turns it into milk)   (milk is food for the person)

Food chains

These links, called **food chains**, show how one organism is eaten by another organism, which can itself be eaten. All food chains start with an organism that can produce its own food – here a green plant. Plants make food by photosynthesis; we call them **producers** (see Topic 2.4). Animals eat, or consume, food, so are called **consumers**. Consumers may be herbivores, carnivores or omnivores. Cows are herbivores because they eat plants. Cats are carnivores because they eat animals. Animals, such as humans, which eat both, are called omnivores (see Topic 1.6).

On the farm

Some of the food chains in the picture are shown on the opposite page, including whether organisms are producers or consumers.

Plants on a farm make the food that feeds the animals that live there, including humans

THE ENVIRONMENT

grass ⟶ horse

corn ⟶ chicken ⟶ fox

pondweed ⟶ tadpole ⟶ fish ⟶ heron

rosebush ⟶ greenfly ⟶ ladybird ⟶ blackbird ⟶ sparrowhawk

producer   *consumers herbivore*   *1st carnivore*   *2nd carnivore*   *3rd carnivore*

★ THINGS TO DO

1 A common meal might consist of burger and fries with tomato sauce. Here's where it comes from:
burger, made from beef (cows, which have eaten grass)
bun, made from flour (from wheat plants)
fries, made from potato plants
sauce, made from tomato plants
a) Write four food chains for this meal.
b) Would you be a herbivore, a carnivore or an omnivore if you ate this? Explain your answer.

2 The diagram shows living things that can be found on an allotment. (Each animal's food is shown in its mouth.)
a) Use the information to write four different food chains.
b) Combine the chains into a food web. Add any other feeding links you can think of for organisms on an allotment.

3 See who can design the biggest food chain in your class.

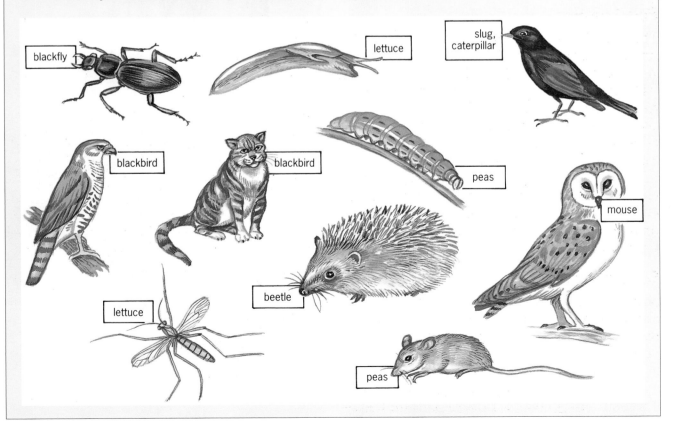

3.7 Food webs

Food from an oak

This oak tree is thought to be the resting place of Robin Hood. It is also a home to a surprisingly large community of animals

The oak tree offers different foods for herbivores, which are then eaten by other animals. A complex web of links can be built up – a **food web**. This shows how many food chains in a habitat can interlink.

Each of the different habitats in an ecosystem will have different communities of organisms living in it, so will have a different food web.

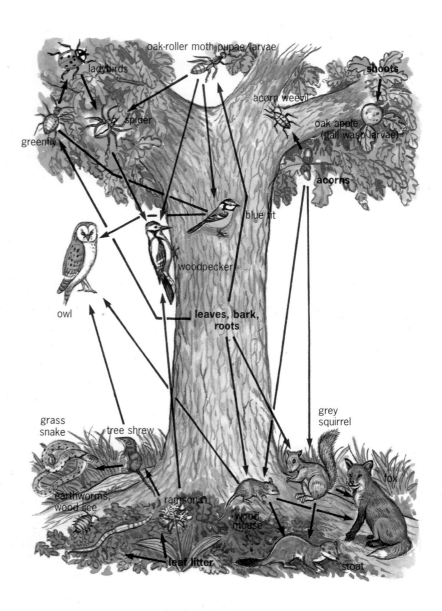

Oak tree food web

Pyramids

A food chain and a food web show how food passes from organism to organism. As this happens, energy and chemicals (nutrients) are transferred within the food. This transfer of food through the food chain helps to keep the organisms alive.

The food chains in the diagram show how three organisms in the photograph are linked together. If we counted the number of organisms at each stage of the chain, we would get the results shown in (a) opposite.

THE ENVIRONMENT

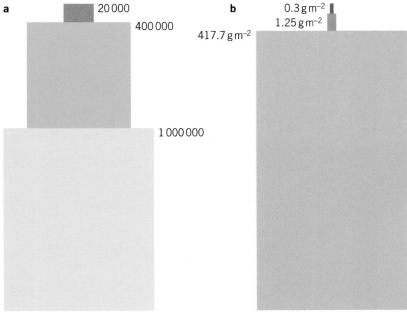

Organisms in a food chain on a savannah, and their pyramids of:
a numbers;
b mass

The numbers of each type of organism can be shown as small squares on graph paper. Drawing around all the squares for each organism gives a box that represents the total number of each type of organism in that ecosystem. The overall shape of the diagram is a pyramid. So it is called a **pyramid of numbers** ((a) in the diagram). A pyramid of numbers shows that there are usually fewer and fewer organisms at each stage of a food chain.

If the mass of each organism is worked out, giving the total mass of each type of organism, at each stage then we can also draw a **pyramid of mass** (b). A pyramid of mass shows that the total mass of organisms gets smaller at each stage of a food chain.

★ THINGS TO DO

1 Greenfly reproduce quickly in warm weather. Suppose all the greenfly on one plant were collected and counted. The results might be as shown in the table.

	Pyramid of numbers	Pyramid of mass
blackbird	1	1
ladybird	40	8
greenfly	700	36
rose plant	1	640

a) Draw a pyramid of numbers using the information. Explain what the shape tells you.

b) Now draw a pyramid of mass. Explain why it is a different shape.

2 Look at the oak tree food web.
a) How many food chains contain five organisms?
b) Write out one food chain with five organisms.
c) Name two different herbivores.
d) Where does the oak tree get the energy it needs to stay alive?
e) If the tree were chopped down, what would happen to the herbivores?
f) How would other living things be affected?
g) What else could affect the food web?

3.8 Energy transfer

Energy conversion

Energy for all living things comes from the Sun, but they are not very efficient at passing it on. A plant converts only 0.01% of the energy in sunlight that falls on it. This energy is stored as food, and then passed on to other substances. But only about 10% of that energy actually passes into the animals feeding on the plant. Less and less energy passes along the food chain, as the diagram shows.

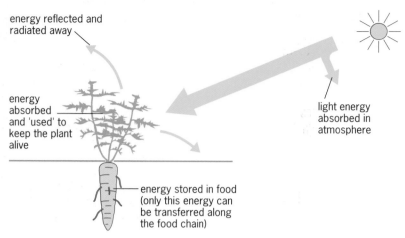

Losses in food production

Energy is not being lost, however, but transferred as each organism respires, grows, makes complex chemicals, and then releases wastes. Living things convert the energy from food into different forms. In warm-blooded animals, much is turned into heat to keep the internal body temperature constant. Mammals and birds need more energy than those animals that do not maintain a warm body temperature. Energy is also converted into sound energy (when shouting or speaking), kinetic energy (when moving) and potential energy (when lifting things).

This pattern of transfer and loss also happens to chemicals in food. Animals at the end of a food chain have fewer chemicals to use, so need more food.

Less and less energy and fewer chemicals are available at each stage of a food chain. The stages, or 'trophic levels', are often limited to four because of the 'losses' along the food chain.

Being more efficient

The amount of food made by a plant during photosynthesis depends on how much light is available, the temperature, the amount of carbon dioxide and the growing conditions in that area. The amount of carbon dioxide in air is only 0.03%. On a bright, hot summer day many plants cannot absorb enough carbon dioxide to maximise food production. Some plants, called 'C4 plants', are naturally more efficient at using it, so they make more food and grow more quickly. Genetic engineering and selective breeding are also now being used to make crop plants better producers of food (see Topics 4.10, 4.17).

The rate at which energy is stored in food made by plants growing in an area is called the **gross primary productivity**. This is a measure of how efficient plants are at storing energy in food. The table shows how plants in different habitats compare.

Habitat	Gross primary productivity/ $kJ\,m^{-2}\,d^{-1}$
desert	<0.5
coniferous forest	2.5
lowland British farm fields	8
tropical sugar cane plantation	24

Minimising energy losses

Eating food from an animal source, such as meat, milk, or eggs, is a less efficient way of obtaining energy, or chemicals. On simple energy-efficiency terms it would be more productive for humans to eat plant rather than animal foods. It is estimated that 11% of the Earth's surface is used to grow plants for food, but about 25% of the land is used to keep animals for food. Some people

THE ENVIRONMENT

argue that this is not an effective way of producing food around the world. They also point out that about 40% of the grains harvested from cereal plants like maize are fed to animals instead of being eaten by humans.

A question of taste?

The article on the right describes battery methods used to rear chickens for meat.

Most people enjoy eating meat, eggs and dairy products; farmers provide these foods by using either traditional 'free range' methods or intensive husbandry

Intensive methods satisfy the needs of the market place, providing enough chicken for people to eat at a cheap price. Cheap food is often the result of practices that have a negative influence on the wider environment and on the welfare of animals. If we demand more care for the environment, and the welfare of animals grown for food, then we would have to pay more. Would people agree to this? Would farmers switch to more environmentally friendly ways if consumers were only prepared to buy foods produced this way?

Following tradition by allowing chickens to roam freely on the farm is the essence of 'free range' methods. Chickens have space to move around as they please (often 350 birds per hectare). Pelleted food is used to supplement the diet of grains and other plant material that the chickens find as they search around in the day. Small huts provide shelter from poor weather. Occasionally birds are killed and eaten by foxes, a natural predator. It takes longer for chickens to be ready for the market. They also cost more, but are said to taste better.

Broilers are ready for the market only 42–47 days after hatching from eggs, when raised by intensive methods. One-day-old chicks are put into single-storey wooden huts called 'broiler houses'. Each hut provides optimal conditions, for the chicks to grow quickly at minimum cost. Water and food are provided by automatic feeders in the warm and well-ventilated artificial environment, often under computer control. The floor is covered with materials like wood shavings to absorb the chickens' droppings. Artificial lighting is kept dim to subdue the chickens and avoid fighting. Each hut will house hundreds of chickens, all the same age. There may be ten birds per square metre of floor space. The health of birds in the hut is regularly monitored, and any dead birds quickly removed to reduce disease.

Under these conditions the chickens gain weight rapidly. Their diet is adjusted as they grow to provide more protein. As a result their muscles may grow disproportionately large, making walking difficult. Droppings from the hut floor are removed and must be disposed of without causing health hazards. This chicken manure is smelly, but can be used to improve soil fertility for growing crops. Or it can be burned to generate electricity.

★ THINGS TO DO

1 a) Explain how plants are inefficient converters of energy from sunlight.
b) Why is producing food from plants more efficient than from animals?
c) 'We could provide more food more cheaply by growing crops like sugar cane.' Write reasons for and against this point of view.

2 a) Make a list of advantages and disadvantages of rearing chickens for meat using intensive and free range methods.
b) Debate with a group of friends the rights and wrongs of each method.
c) Write a code of animal welfare for farmers to follow to safeguard the rights of animals reared for food.
d) Find out how intensive and free range methods are used for egg-laying hens.

3.9 Community relations

Dependence
Plants and animals living in the same area form a mixed community. They may depend on each other for their survival, as the photo on the right shows.

Competition
Plants and animals may also compete. For example, plants need water, air, nutrients from the soil and sunlight to survive. Plants growing together compete for these. Smaller plants may be so shaded by larger ones that they do not get enough sunlight to produce their food, so will die. This is why gardeners space their seeds when sowing, and thin out the seedlings.

Animals also compete against one another. The woodland and forests of Britain are the home of two types (species) of squirrel, the red and grey. Both have the same needs, but the grey squirrel is bigger, and a less fussy eater and so at an advantage. If both squirrels coexist in a forest it is not long before the red squirrel population declines. The map shows how the populations of red squirrels have declined since 1940.

Insects fly from flower to flower in search of nectar (their food). As they do this, they carry pollen from one flower to another (**pollination**). Only when flowers have been pollinated can they produce seeds, which will ensure their survival in the following years. So both insects and plants benefit

The red squirrel is a native species – it has lived in British forests for thousands of years. But you are more likely to see grey squirrels. These were introduced to Britain from America

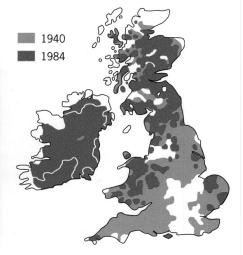

Population of red squirrels in Britain, 1940 and 1984

Predators and prey
Within any community of plants and animals, the population of some organisms may be increasing whilst others are decreasing. For instance, if there are lots of foxes (**predators**) in an area, the rabbit population (the **prey**) will quickly decrease. If there is insufficient food to maintain all the foxes, some will die unless they move to another area where there is sufficient food, or eat other things. In recent years foxes have begun to move into city areas to find food.

Foxes are predators; they rely on smaller mammals, such as rabbits, for food

THE ENVIRONMENT

The same thing happens with insect populations. A mild winter and a warm spring can mean lots of greenfly in the garden. A rise in their population means more food for animals that eat them, such as ladybirds. So their numbers also increase.

Biological pest control

The warm, dry conditions in a greenhouse are ideal breeding grounds for a number of pests, which weaken the plants so they become prone to disease. In the past, gardeners used chemical sprays to kill these pests. However, the sprays not only taint crops such as tomatoes, but also kill harmless insects. By introducing natural predators, such sprays can be avoided. A predatory type of spider mite called *Phytosieulius* eats up to 20 red spider mites every day, and so steadily eliminates them. The effect is called **biological control**.

Red spider mites feed on the sap of greenhouse plants, causing leaves to turn brown and drop off

★ THINGS TO DO

1 The table shows the different plants growing in a lawn, surveyed with a quadrat.

Plant	% of lawn covered
grass	65
clover	10
moss	8
dandelion	11
buttercup	6

a) Draw a pie chart to show the percentage coverage by each plant.
b) Estimate what the figures could be in 12 months' time if the weeds were not pulled out.
c) How could you test the difference between leaving the weeds alone and removing them?
d) Explain why weeds, with their wide, long leaves and long roots, might eventually outgrow the grass plants.

2 The graph shows how the populations of predators (ladybirds) and prey (greenfly) change over 3 years.
a) Why will there be more ladybirds when there are more greenfly?

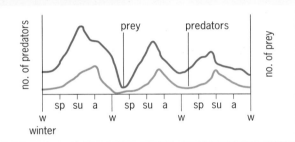

b) More greenfly die when winter is very cold. Which year do you think had the coldest winter? Explain your answer.
c) Draw your own graph showing what you think would happen if we had two very mild winters, one after the other.

3 Use the information on biological control to answer the following questions.
a) Describe how the number of red spider mites can be reduced using biological control. Use the words 'predator' and 'prey' in your answer.
b) Why would this method be less effective outside a greenhouse?
c) Find some other examples of biological control, including the use of myxomatosis to control rabbits in Australia.

3.10 Dynamic numbers

The size of any population of plants or animals changes with time; it is 'dynamic'. The size is influenced by factors such as:

- competition for food or nutrients,
- competition for space or light,
- the amount of grazing of plants,
- the amount of predation,
- the spread of disease,
- the rate of reproduction,
- the rate of death,
- the rate of emigration and immigration,
- the effect of damage by humans.

Year	Number of seals on one island/1000s
1960	8
1965	24
1970	47
1975	95
1980	167
1985	195
1990	309

Changing numbers

Populations of fish (predators), herbivores (prey) and plants in a lake

The graph shows how populations of fish (predators), various herbivores (their prey) and plants in a lake may vary over a year. In early spring the plant population increases as more light is available and the water gets warmer. The individual plants start competing vigorously with each other for space, light and nutrients. Meanwhile they are being eaten by herbivores such as insects and crustaceans, which also compete with each other; the predators likewise also compete among themselves. The size of the fish population depends on the combined effect of all these factors. There can be a large fish population only if there is sufficient food for them. If any of the other factors change, then the number of fish also changes.

Krill is the main diet of seals

Seals and fishing

A survey of the Antarctic fur seal (*Arctocephalus gazella*) in the south Atlantic (shown in the table) shows a dramatic rise in population. This is particularly significant since it was thought that sealers from Britain and America had hunted the seals to extinction by 1830. The increase in numbers is thought to be due to a ban on hunting them and an abundance of their main food source in the sea – shrimp-like animals called krill. The reduction of whale populations in the area has helped since whales compete for the same food.

Similar increases in seal populations off the coast of Canada are worrying fishermen. They blame seals for causing a fall in fish stocks. They claim that seals eat fish and demand that seals are killed to control their numbers.

THE ENVIRONMENT

More and more

The table and graph show that the world's population is growing rapidly. The rate of increase is greatest in countries in Africa, Asia and South America. Population growth depends on both birth and death rates: large families may reflect social structures and religion, also improved food and water availablility and better health, child care and treatment of disease so people live longer. Wars and famines have dramatic effects on population – about 55 million people were killed in 6 years during World War II.

Country	Population/millions 1990	2025	Total fertility rate
China	1134	1569	2.5
India	850	1365	4.0
United States	250	319	1.9
Indonesia	178	265	3.1
Brazil	150	224	3.3
Russian Federation	148	153	2.3
Japan	124	126	1.6
Pakistan	112	244	5.9
Bangladesh	107	180	4.6
Nigeria	96	217	6.0

Growth rates in the world's 10 most populous countries (A fertility rate of 2.0 means that a couple will produce two children, so the population will not grow. A rate above 2 indicates a growing population, as long as children survive to reproduce.) (from *Science and Public Affairs*, winter 1994, Royal Society/BAAS, p. 10, with kind permission)

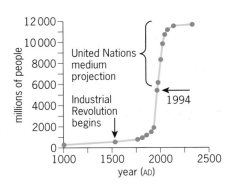

World population changes since 0 AD (from *Science and Public Affairs*, summer 1994, Royal Society/BAAS, p. 41, with kind permission)

★ THINGS TO DO

1 The graph shows the effect of overfishing on the fish population in a lake.
   a) Copy the graph into your notebook.
   b) Describe what is happening and why.

   c) Draw sketch graphs showing what might happen to the numbers of herbivores (prey) and plants in the lake. Add a brief explanation.
   d) If a disease killed 30% of the herbivores, how would the numbers of fish and plants change?

2 Look at the information about seals opposite.
   a) Seals can swim from island to island. Draw a graph of the data and then show how seal immigration (arrivals) and emigration (departures) would affect their population.
   b) List arguments for and against killing seals.
   c) A ban on killing whales would result in more whales. How might this affect seal numbers?

3 a) What is the estimated human population for the year 2000?
   b) Why is the population rise likely to continue?
   c) What evidence suggests that contraception might help to stop this increase?
   d) What factors may halt the increase by the year 2500?

3.11 Managing the environment

The survival of plant and animal communities often depends on humans. Habitats are often destroyed to build houses, factories and roads. Chopping down forests and hedgerows or draining wetlands to provide extra farmland removes valuable homes for wildlife. Converting huge areas of land into mineral quarries threatens areas important to plants and animals. And dumping waste from factories and homes pollutes and destroys habitats.

But some human activities are environmentally friendly. Good planning and management can help retain much of what is needed in the habitat for our animal and plant neighbours to survive.

The following letters were written by two young people on work experience in places that try to improve the environment.

New roads may cut through areas where wildlife would otherwise thrive

Ranji: forestry

I've learned lots about caring for the environment here and now know what hard work it is being a forestry ranger! The forest is managed for three uses.

The first is to grow trees for timber. Britain grows only about 10% of the trees it needs, and most of these are conifers which don't lose their leaves in winter, so keep on growing. They reach their full height in about 40 years, growing in soil that is too poor for food crops. Conifers grow tall and straight, so are ideal as timber.

The second use is leisure and tourism. People come to picnic, talk, ride bikes and horses and see the wildlife.

That's the third use, of course - to provide different habitats for wildlife. I've been helping to plant trees in areas where conifers were cut down. We've planted oak, ash, elm and maple trees. These are deciduous trees (lose their leaves in winter) so grow slower and are less useful for timber. They will soon be the home of a different type of plants and animals to those which live in the conifers. Wildlife conservation really works here.

Jeff: wildlife action group

My job has been really great. Our action group helps schools to create new habitats, mostly wildlife areas and ponds. I've been visiting primary schools, digging out and laying artificial ponds. I hadn't known before, but the number of amphibians in Britain has dropped markedly over the past 50 years. It's strange, but true, to think that, unless action is taken, frogs, toads and newts are on the road to extinction! The main problem is that amphibians need water to breed in. It's estimated that there are 5% fewer ponds compared with 10 years ago. Most have been filled in or drained because the land is more valuable for building or farming.

Within a few years this pond I photographed will be home to water plants, amphibians, fish and other aquatic animals. What's more, it will attract lots of different birds and other animals. It's an extra habitat and one that's much needed.

I'm hoping to go back next year to learn how to plant hedges. A hedgerow is another valuable habitat for plants and animals.

★ THINGS TO DO

1 List eight different ways in which humans make it difficult for wildlife to survive.

2 The photograph shows how one council is conserving wildlife living alongside a country lane by not cutting the verge and hedgerow.

a) What different habitats can you see?
b) Why would a proposed plan to widen the road be a bad idea here?
c) Write an article for a local newspaper suggesting what can be done to conserve places for wildlife.
d) Find out from your local council what it is doing to manage the environment.

3 The way that farmers use their land continues to change. The table shows how hedgerows have been removed to make bigger fields, which are easier to harvest.

Year	Total length of hedgerows in England and Wales/km
1947	792 000
1969	699 200
1980	649 600
1985	617 000

Surveys show that populations of wild animals and birds living on farmland have altered as a result.

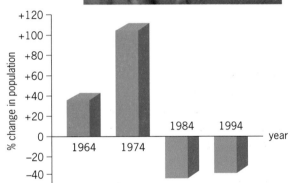

The graph shows how the populations of a common bird, the bullfinch, have changed (relative to a 1980 baseline of 0).

a) Draw a line graph to show changes in the amount of hedgerows.
b) Describe how and why changes have occurred.
c) How might these changes affect animals, like birds, living on farmland?
d) What further information is needed to prevent changes in farming endangering wildlife?

3.12 Decay for renewal

We all know that food that is kept too long begins to decay – it goes off. Some of the bacteria that cause food decay can be dangerous, causing severe food poisoning, fever and diarrhoea. Sometimes we cannot see that the food has gone off, but the bacteria are still there.

Decay is important for the survival of plants and animals in an ecosystem. As plants grow they absorb nutrients (minerals) from soil. When animals eat the plants they obtain the nutrients they need. When dead plants and animals and their wastes decay these nutrients are released back into soil. Decay recycles the nutrients needed for life to continue.

Decay occurs in almost all habitats. During this process, dead organisms, wastes and food are digested by microbes, breaking down large molecules into smaller ones. Materials that can decay are called **biodegradable**. All natural materials are biodegradable. Man-made materials, like glass and plastic, are usually not.

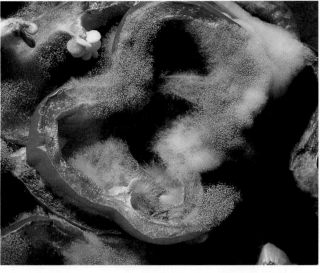

Fruit and vegetables like this pepper decay in a spectacular manner, becoming mushy and covered with fungi (mould)

Stages of decay

Microbes (bacteria and fungi) work best in warm, moist places (such as ponds), with enough air for respiration. As decay happens nutrients are released, including nitrates, phosphates and potassium. Two gases are produced: carbon dioxide and methane. In small ponds, the methane produced by the decay of leaves falling in during autumn can kill small animals and fish living there.

The different stages of decay of life in a pond are shown in the diagram. The microbes that cause decay are found in the dead fish, leaves and mud at the bottom.

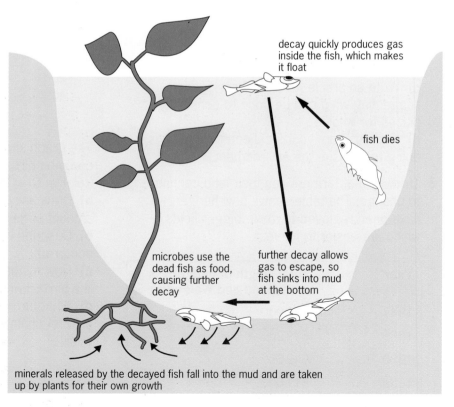

Death and decay in a pond

What a mess!

Cow pats in a field may look unsightly and smell unpleasant, but not for long

A cow pat offers food to a small community of microbes and animals. These organisms start to change it, and decay begins. Within a few months the cow pat will have been broken up, but acid released during decay will prevent plants from regrowing. Later, after rain has washed away the acid, plants will begin to grow once again. Minerals from the decayed cow pat will encourage good plant growth. The cow pat will eventually disappear.

The age of a cow pat can be found by looking at which organisms are feeding on it. The table and diagram show different organisms found at different ages.

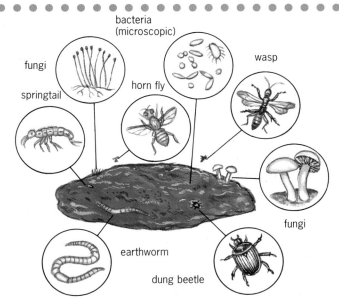

Organisms that may be found on a cow pat

Age of cowpat/ weeks	Common organisms
1–2	flies, beetles, wasps
2–6	bacteria, fungi, flies, beetles, wasps, springtails
5–16	bacteria, fungi, flies, beetles, wasps, earthworms

★ THINGS TO DO

1 Growing microbes on decaying food can be a health hazard. There is a risk that some microbes might cause disease. But you can grow fungi quite safely as follows. (Check with your teacher first though.)

1. Find a jam jar or sweets jar, wash it out and dry it.
2. Add some sand to cover the bottom.
3. Pour water in to moisten the sand.
4. Place some old bread or fruit on the wet sand.
5. Cover the jar with a lid that has a hole. Place cotton wool in the hole to block it lightly.
6. Leave your jar in a safe place where it will not be disturbed.
7. Observe and draw any microbes that grow (you are most likely to see fungi). Do *not* open the jar or spill anything from it. Wash your hands afterwards.

This method supplies moisture, which microbes need to grow.

a) What will be their food?
b) What difference might there be from leaving the jar in different temperatures?
c) How will the cotton wool prevent microbes from escaping?
d) Why is this important?
e) Why is it important that the microbes can get air through the cotton wool?
f) Make a list of things that could affect how quickly bread or fruit decays. Plan how you could use the method described above to test your ideas. Carry out tests to find out which of the factors seems most important.

2 a) What are the advantages of organic materials decaying?
b) What are the disadvantages?

3.13 Using decay

For a 12-month trial period, Townshire Council is offering free compost makers to some householders. It wants to know if people will use composting to get rid of their biodegradable wastes, instead of throwing them away. It is trialling three types of compost maker to find the best method.

Max has an insulated plastic bin. It is compact but can make up to three times the amount of compost as a simple compost heap. The plastic absorbs heat from sunlight and stays warm. This helps the microbes inside to be active and decay the household and garden waste added to the bin daily

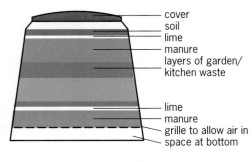

Jim chose a tumbler. It is about the same size as Max's, but is held on a strong metal frame, allowing Jim to rock the tumbler each day. This makes the microbes inside more active by mixing in more air. It is especially effective with kitchen waste and grass cuttings. Compost can be made in only 21 days

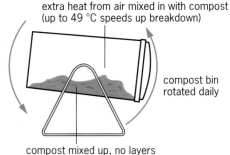

Ben chose the most expensive version – a wormery. This is more complicated, but smaller than the others. 'Tiger worms' inside help to decay kitchen waste and dead plants. The compost can be used as a fertiliser, for growing seeds or potted plants and to improve soil. A tap at the bottom allows liquid to be removed; this can be used to 'feed' houseplants

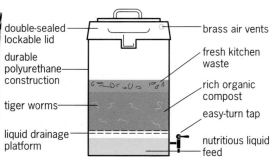

More from cow pats!

Cow dung, the material that forms cow pats, can be a useful source of energy. In countries like India it is dried in the Sun and then used as a fuel. You might be surprised at the amount of heat released when it burns! An alternative way to use dung is to let it decay in a giant container. The methane gas released during decay can be collected and piped into a boiler to be burned.

Methane released from decaying waste isn't always useful. Landfill tips, where household rubbish is dumped, contain much organic waste. This is covered by soil to stop animals like mice and rats breeding and spreading disease. But any methane released by decay could build up in the soil and cause an explosion. To prevent this happening, pipes are placed in the soil to allow any methane to escape into the air.

THE ENVIRONMENT

Treating sewage

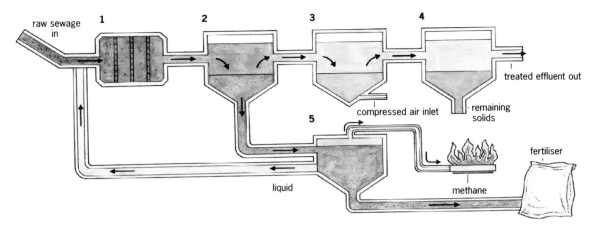

1. solid objects, including rags and paper, are filtered out
2. primary settlement tank: heavy solid lumps settle to the bottom of the tank and later pass to an anaerobic digester
3. anaerobic digester: microbes decay the solids without air, releasing methane and leaving sludge, which can be used as a fertiliser
4. aerobic treatment (filter bed): the solution is sprayed over chunks of coke which are covered with microbes; this provides a large surface for mixing sewage and microbes and plenty of air to keep the microbes active — sewage is digested by microbes and harmful bacteria are killed
5. final settlement tank: any remaining solids settle out of the water, which is now safe enough to be released into a river or the sea

Stages of sewage treatment

★ THINGS TO DO

RISK

1. Make a list of factors that could affect the rate at which grass or food decays. Write a plan to test your ideas, using grass cuttings or leaves. Mention any precautions to take against the spread of harmful microbes.
   After completing your tests, prepare a short article for a gardening magazine on how gardeners can prepare compost most rapidly.

2. The average household dustbin contains the following waste each week.

paper and cardboard	30%
plastics	8%
metals	8%
vegetables and other food	25%
clothing materials	4%
glass	10%
other rubbish	15%

a) Turn the data into a pie chart or bar graph.
b) Which materials could be composted?
c) How would the council benefit if everyone used composters?
d) How would householders benefit?
e) Which materials could be recycled instead of being thrown away as rubbish?
f) Draw a poster to encourage people to make compost with their kitchen and garden wastes and to recycle non-biodegradable materials.

3. Shellfish like cockles and mussels filter their food from the sea. In coastal towns raw sewage is often piped straight into the sea, where the water currents dilute and carry it away. Microbes in the sewage are not killed, however, and may remain in the water for shallow-water creatures to take in. A sewage treatment plant might help solve some problems caused by this method of disposal.
a) Explain why a sewage treatment plant might not solve *all* of the above problems.
b) Describe the advantages and disadvantages of building a sewage treatment plant.

3.14 Round and round

Nutrients are recycled within the ecosystem, so they are available continually for all living things. This recycling is vital for life to continue. Carbon is a vital chemical element for all living things. It is combined with hydrogen and oxygen in molecules of carbohydrates, fats and proteins. Animals cannot use pure carbon, but rely on plants 'fixing' it (turning it into another more usable form). Plants do this when they take in carbon dioxide during photosynthesis, and convert the carbon into carbohydrate molecules (sugar and starch). The carbon can then be further converted into other molecules, including oils and proteins. These substances are solids or liquids, so stay inside the plant. Any excess unfixed carbon remains as carbon dioxide gas, and escapes through the stomata (see Topics 2.4, 2.5).

Animals along a food chain obtain their carbon from their food. Herbivores, for example, obtain it as carbohydrates and starch in their plant diet. The carbon is then 'unfixed' by being changed into carbon dioxide and escaping into the air. All living things release carbon dioxide during respiration; it is also a feature of decay.

In a stable community there is a constant amount of carbon being recycled. So the amount of carbon dioxide removed from air by plants equals that released in respiration.

uptake of carbon dioxide = release of carbon dioxide

The amount of carbon dioxide in air remains constantly at about 0.03% – a very small amount for something so important.

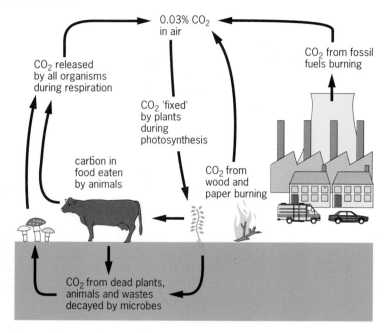

The carbon cycle

Carbon dioxide levels in the atmosphere have risen from 280 to 350 parts per million (p.p.m.) in the last 100 years

New evidence emerges that icecaps at the Earth's poles are melting and sea levels rising as global temperatures rise

Fears that more cars, lorries, and factories in the future will lead to global warming

Greenhouse gases like water vapour, carbon dioxide and methane are trapping more heat in the atmosphere

THE ENVIRONMENT

When fuels (e.g. petrol, gas) are burned they release carbon dioxide into the air. Animals release carbon dioxide into the air. Draining peat bogs does the same. Chopping down forests means that less carbon dioxide is removed from the air (and less oxygen is released into it). Scientists are worried that carbon dioxide is building up in the atmosphere and causing an increase in the **greenhouse effect**, which in turn may increase 'global warming'.

The decline of British peat bogs

Many peat bogs are now being drained and the land used to develop coniferous forests, as peat is less valuable than wood. The wisdom of draining peat bogs to plant coniferous forests is now being questioned, however. First, the habitats of many species of plant and animal are destroyed. There are also other dangers.

In Britain, more carbon is stored in soil than in plants. Soil contains carbon in the form of humus (dead plants and animals plus organic waste). Although peat is a poor type of soil, Scottish peat bogs hold 75% of all the carbon stored in Britain. It would take 100 years of burning fossil fuels in power stations and vehicles to release this amount of carbon into the atmosphere. But destroying peat bogs will release the carbon stored in peat into the air. If water is drained from a peat bog, carbon dioxide bubbles out. Decaying dried peat releases more gas. It would be equivalent to chopping down tropical rain forests. The overall result is an increase in the amount of CO_2 in the atmosphere and so in the greenhouse effect.

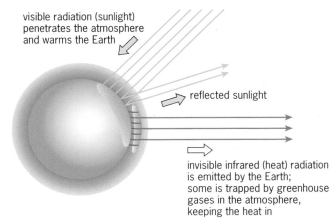

How the 'greenhouse effect' occurs

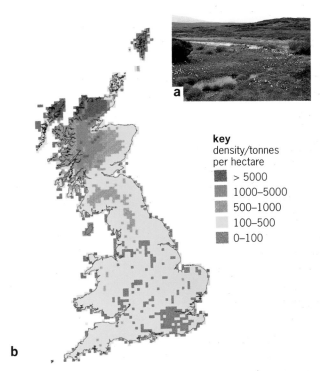

a A peat bog; **b** peat bogs in Britain (from *New Scientist* 19 November 1994, p. 6, with kind permission)

★ THINGS TO DO

1 Copy the diagram of the carbon cycle into your notebook and explain:
   a) how CO_2 is made available for plants,
   b) how animals obtain their carbon,
   c) how microbes obtain their carbon.

2 Global warming may result from 'greenhouse gases' accumulating.

   a) What gases cause this effect?
   b) Explain four different human activities that can contribute to global warming.
   c) What could be done to reduce their impact?
   d) Some scientists disagree that all is bad, saying that plants will benefit from global warming effects. How do you think they could benefit?

3.15 Nitrogen for protein

Plants convert nutrients in fertilisers into the complex chemicals they need. For instance, plants need to make nitrogen-containing protein to stay alive and grow. However, they cannot use nitrogen from the air directly for this. Nitrates in fertilisers supply the nitrogen in the form they need.

Each time a lawn is cut nitrogen will be removed as chemicals in the cut grass. Some gardeners therefore compost their grass cuttings, scattering the compost with its nutrients back into the soil. Animals also remove nitrogen when they eat grass, and more is lost as rain water washes nitrates deeper into the soil. As time passes, the grass plants will have fewer nitrates (and other minerals) to use.

But the plants will not run out of nitrate entirely because of the **nitrogen cycle**. The diagram shows how nitrogen is recycled between different organisms within an ecosystem. In a stable ecosystem, the amount of nitrate removed from the soil is usually balanced by the amount replaced in other ways.

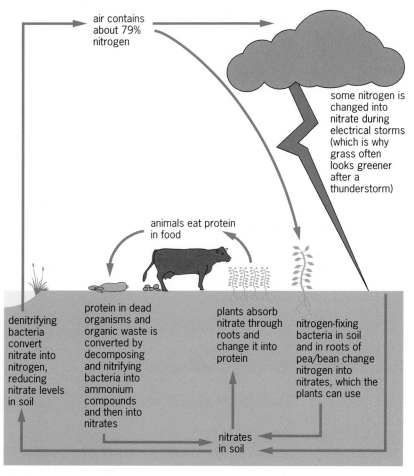

The nitrogen cycle

Adding nitrogen, losing nitrogen

Arable farmers often use large amounts of fertiliser to improve crop yields. However, it is important to apply it only when plants are growing and actually needing it. This is because any nitrate or phosphate that remains in the soil may be washed out by rain water into ponds or rivers, causing **eutrophication**. This is overgrowth of algae in the water because of over-enrichment with nutrients. Slurry (animal droppings) also pollutes water in the same way if accidentally spilt into streams. High levels of nitrate in drinking water (above 50 mg l^{-1}) are harmful to our health, and may be a factor in stomach cancer and 'blue baby syndrome' in young babies.

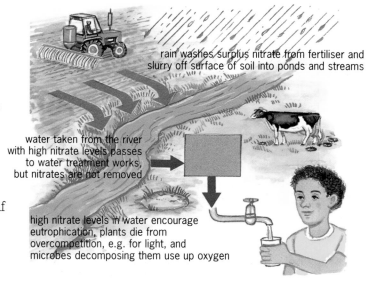

The story of nitrate pollution

THE ENVIRONMENT

A pond showing eutrophication; there isn't enough dissolved oxygen for fish and other animals to survive, and the pond has little life in it

★ THINGS TO DO

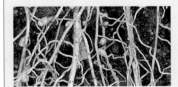

Root nodules of a bean plant, and the nitrifying bacteria they contain

1 The photograph shows the roots of bean plants; the lumps (nodules) on the roots contain millions of nitrogen-fixing bacteria, which make nitrates. After beans have been produced and harvested, the plants will be dug into the ground so that they decay.
   a) Give two reasons why the next year's crop will benefit from the bean plants being dug into the soil.
   b) Clover plants also have root nodules. Why will these weeds grow quickly in a lawn, even if no fertiliser is used?

2 The graph shows how much nitrogen is needed by a field of wheat plants and how much is available from the soil.

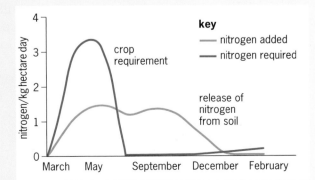

a) How much extra nitrogen is needed by the wheat plants in May?
b) Why does the amount of nitrogen needed drop so much by August?
c) New wheat seedlings grow slowly over the wet winter months. What will happen to surplus nitrate during this time?
d) What problems could this cause?

3 The table shows how much nitrate we get from food.

water	20%
vegetables	55%
milk	18%
cheese	2%
meat	5%

a) Draw a pie chart to show this data.
b) Why do vegetables contain so much nitrate?
c) Why does food from animals contain less nitrate?
d) Why is water taken from rivers in lowland areas more likely to have higher nitrate levels than water from upland streams?
e) Contact your local water company for information on how water nitrate levels are monitored.

3.16 Pesticides

Types of pesticide

Pests cause damage to garden plants. They can destroy the leaves (so the plant cannot make enough food), spread diseases and spoil vegetables. Pesticides are chemicals that are used to destroy pests before they can cause damage. General pesticides can kill a wide range of plants or animals. Pesticides can also be selective, and designed to kill specific plants (see Topic 2.15) or animals, as the illustration shows. Chemicals used in the garden include: herbicides, which kill plants, insecticides, which kill insects, and molluscicides, which kill slugs.

Pest	Problems caused	Pesticide				
		Rapid	Sybol	Picket	Bug gun	Slug xtra
capsid bug	attacks herbaceous plants like roses, dahlias – small holes in leaves, lopsided flowers		*	*		
earwig	affects dahlias, chrysanthemums – chews holes in leaves and petals		*	*		
flea beetle	small round holes in leaves of wallflowers		*			
leaf hopper	leaves of roses turn white/yellow		*	*	*	
slug	eats most flowering plants – irregular holes in leaves, stems eaten at ground level					*
wireworm	feeds on roots and tubers of most flowering plants		*			
woolly aphid	small white furry clusters under leaves of shrubs and trees	*				

Problems with pesticides

On farmland, pests can significantly reduce the size of a crop, reducing the food supply. By cutting down on the number of pests with pesticide use, we can increase the crop yields and meet the needs of the consumer better.

When any pesticide is applied, care is needed. The pesticide should be restricted to areas where it is needed, or other plants and animals may be killed accidentally. The correct amount should be used: too little will not have enough effect; too much may leave surplus pesticide, which could continue to kill plants or animals.

Many pesticides break down quickly after being used. This means that their ability to kill pests is short lived, so they may have to be applied repeatedly. Persistent pesticides have the advantage of lasting for a longer time, but this can have harmful as well as beneficial effects.

THE ENVIRONMENT

Effects on wildlife

Pesticide use can have disastrous results, as research in the 1950s revealed. A sudden drop in the population of sparrowhawks in Britain was linked to high levels of an insecticide called DDT in the dead birds' bodies. It was also found in their eggs. In fact, the egg shells were much thinner than normal, so thin that some had been smashed, killing the chick inside. The DDT was traced back through the sparrowhawk's food chain to insects being killed in nearby fields. It is now known that DDT cannot be excreted by animals that eat it. By eating prey that had DDT in their bodies, the sparrowhawks were slowly accumulating the pesticide in their own bodies until it started to poison them.

and eels that have absorbed the chemicals from the water. Fortunately, the most harmful pesticides are now banned in Britain and other countries. The populations of sparrowhawks and otters have recovered as a result.

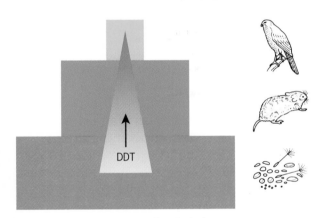

DDT accumulation in the food chain

Other pesticides have similar effects. Otters have disappeared from most areas of Britain where they once lived. Their breeding success is known to be harmed by organophosphate pesticides like dieldrin, which build up in their bodies as they eat fish

Both sparrowhawk and otter populations have suffered from pesticides in the food chain

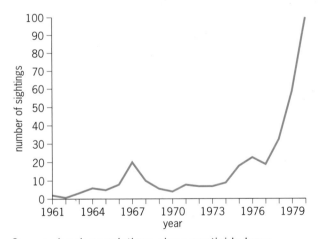

Sparrowhawk populations since pesticide bans

★ THINGS TO DO

1 Look at the data on different pesticides.
   a) Why can Rapid be called a selective pesticide?
   b) What problems would be cured by spraying roses with Sybol?
   c) Dahlias have underground tubers. How could they be affected and treated?

2 New pesticides are being developed by chemical manufacturers.
   a) Write a list of properties that you think these new products should have (advantages).
   b) Write another list of properties they should *not* have (disadvantages).

3.17 The organic way

Curt doesn't use artificial chemicals such as fertilisers and pesticides to grow plants – he is an organic grower. He explains why, and some of his methods.

I TURNED TO ORGANIC GARDENING AFTER MY FARMER NEIGHBOUR FRANK TOLD ME HE SPENDS A FORTUNE ON FERTILISERS AND PESTICIDES. HE SPRAYS HIS WITH PESTICIDES UP TO FIVE TIMES AS A CROP GROWS. I WONDER HOW MUCH OF THE CHEMICALS ARE LEFT IN THE FOOD WE EAT, AND WHAT HARM THEY MIGHT DO. HOW CAN HE PREVENT PESTICIDES BEING BLOWN ON TO THE HEDGES AND ELSEWHERE? THEY COULD BE DAMAGING THE ENVIRONMENT IN UNKNOWN WAYS. SO I DECIDED NOT TO USE ARTIFICIAL CHEMICALS. SOMETIMES MY PLANTS ARE SPOILED BY PESTS AND I DON'T GET HIGH YIELDS. I FOLLOW SIX MAIN RULES.

1. I treat all of my garden in the same way. I grow all of my plants the organic way: fruit and vegetables, trees, shrubs, flowers and lawn. I keep my garden a safe place for people, pets and most wildlife. Harmful rubbish is removed and biodegradable materials are composted.

2. I keep the soil in perfect condition by digging regularly, and adding natural manure to give good soil structure, drainage and nutrients. Where I have a patch of land I'm not using I grow 'green manure' plants. They stop weeds growing and take up nutrients that would otherwise be washed out of soil by rain. When they've grown I dig them straight back into the soil so they decay and release their nutrients. Lupins are a favourite of mine because they're lovely flowers as well. I also grow plants like feverfew, which have enormously long roots and bring up nutrients.

3. I test the pH of my soil and choose plants that grow well in those conditions. A plant that prefers to grow in a dry, sunny place will struggle to survive in the shade. So don't plant it under trees, is my view. Some plants prefer growing in light sandy soils, not the heavy clay soil in my garden. So, I don't grow them.

4. I try to grow a good range of plants for variety. That way, more insects and other animals are attracted into the garden. Some flowers are left to go to seed to attract birds and mammals. I leave some plant material to decay as places for scavenging and decomposing animals. I've also made a pond, which encourages predators like blackbirds, hedgehogs and beetles to visit and eat pests like caterpillars, slugs and millipedes.

Some plants attract beneficial insects into the garden: **a** the 'poached-egg' plant (*Limnanthes douglasii*) attracts hoverflies, which eat pests; **b** the 'butterfly bush' (*Buddleia davidii*)

THE ENVIRONMENT

Plastic barriers against carrot fly

Mulching methods for preventing weed growth

5 I don't use artificial chemicals to control pests and diseases on my plants. I avoid pesticides because they could kill animals like ladybirds, which can help me. These predators are quite effective at keeping down pests like greenfly. They simply eat them.

   I pick off pests like caterpillars and slugs from cabbages. I also catch slugs with traps made from upturned grapefruit/orange peel halves and shallow trays of beer. I cover carrot plants with a clear reusable mesh material or put low plastic barriers up to stop carrot flies laying their eggs and ruining the plants.

6 I put black plastic down on the soil as a mulch. It is great for preventing weeds growing. Mulching with grass cuttings, bark chips or dead leaves is even better because they release nutrients when they decay. If they fail I can always dig out weeds or even burn them with a flame thrower.

★ THINGS TO DO

1 You can find out how many slugs are in an area using the 'slug traps' mentioned in number 5 above.
   a) Test different ways to collect slugs in one area. Which seems to be the most effective?
   b) Use the best method to estimate the number of slugs in different areas. Rank the areas in terms of need for slug control.
   c) How could slugs be controlled without using a pesticide?
   d) What benefits would this bring?

2 Consider the advantages and disadvantages of growing organically. Present your ideas as a poster.

3

Red clover

Plant	Growth period/months	When to sow	Nitrogen fixing?
buckwheat	1–3	Apr.–Aug.	no
field beans	6	Sep.–Nov.	yes
red clover	3+	Apr.–Aug.	yes
lupins	2–4	Mar.–Jun.	yes
radish	6	Aug.–Sep.	no

The table and photo show some types of green manures (plants that are dug into the soil to add nitrogen).
   a) How are nitrogen-fixing green manures different to non-nitrogen-fixing types?
   b) Write a list of plants that would provide green manure throughout the year. Explain your choice.

3.18 Pollution

Our environment can be harmed by chemicals we make and use. This is called pollution. The chemicals can be called **pollutants**. Pollution can also result from excessive noise or light, or careless disposal of wastes. The harmful effects of pollution are not new.

Analysis of ice in Greenland shows 'leaded snow'

Researchers have found lead in cores of ice laid down 2500 years ago. Mining lead ore (galena) and extracting lead from it by strong heating (smelting) started in Greece at this time. Lead was used by Romans up to AD300 for toilets, roofs and water pipes. Mining and smelting were extensive in Germany in the Middle Ages. Lead mines were active in Britain around the start of the 1900s, but have now closed.

Bones excavated from Roman sites show high levels of lead absorbed in the food they ate and dissolved from water pipes made from lead. Lead damages health and can result in poor mental development of young children. It also restricts the growth of most plants. Lead pollution is a problem today, as lead particles are released into the air in traffic fumes.

Lead free

People living near busy roads are becoming increasingly concerned about the amount of lead in the air they breathe and in the soil in which vegetables are grown. Encouraging drivers to use lead-free petrol has reduced the lead content of the air in recent years. The table below shows how the amount of lead in the soil varies at different distances from a busy road.

Distance from road/m	Amount of lead/ mg lead per kg soil
1	980
10	110
25	45
50	33
80	20

Tell-tale signs

Areas where few plants are growing may be polluted. Soil here may be the remains of spoil heaps from old lead or copper mines. Only plants that can tolerate high levels of these substances can grow, so plant types can give some indication of the type of pollution in the soil. The seeds of other plants may germinate but quickly die.

The giant stinging nettle, *Urtica dioica*, thrives on soil that is high in phosphates. It grows well in graveyards in soil where phosphate is released from rotting bones. But it also now grows in areas where phosphates from artificial fertilisers have accumulated in the soil. So finding this plant in places where it would not normally grow is an indication of phosphate pollution

THE ENVIRONMENT

Land reclamation

The sort of reclaiming shown in the illustration improves the appearance of the landscape, but pollution can still occur. Rain water running out from spoil heaps or old mines can carry harmful chemicals. Experiments are being carried out to see if bacteria can remove these chemicals from water before they cause pollution. Sulphur-reducing bacteria can remove heavy metals like cadmium. Others may stop iron being washed from abandoned coal mines and turning rivers acidic.

It's hard to see that this hill was once a spoil heap made of wastes from a coal mine. The waste is still there, underneath soil that was added when the heap was reclaimed. This process is expensive, but recent research shows that cheap materials like ash from power stations and silt dredged from river beds can make a reasonable soil for some plants

Recycling can help

As seen in Topic 3.13, many materials currently thrown away could be recycled and re-used. Advantages of this include:

- less waste is dumped at landfill tips so less land is needed, with less pollution,
- generally, less energy is used to recycle materials than to make them from their raw materials,
- fewer pollutants are produced,
- limited supplies of raw materials are not used up.

Materials such as glass, paper, wool, plastic, iron and aluminium can be recycled

★ THINGS TO DO

1 a) Draw a line graph to show how lead pollution varies at the side of roads.
b) Why should people living near busy roads be worried about lead pollution?
c) What can be done to reduce this problem?
d) Some old houses have lead water pipes. Why should they be replaced by plastic?
e) Adding lead to paint used on children's toys used to be common, but is now banned. Why?
f) Explain why pollution is not just a modern problem.

2 a) Recycling helps to reduce pollution. What materials can be recycled?
b) Survey your friends, class or school to see how many people recycle materials, and which. Display your results.
c) Find out where your local recycling sites are.
d) Write an article or advert to persuade everyone to recycle.

3.19 Pollution control

It is illegal to put engine oil down the sink. The oil would end up in the water system, killing animals and plants, and possibly harming people. So too would other chemicals, including heavy metals like copper and cadmium, printing inks, organic compounds, farm slurry, fertilisers and pesticides. Laws prevent anyone polluting water in rivers, ponds or the sea. But sometimes they are broken, and accidents also happen. Dying fish or birds are often the first signs of water pollution. In fact, animals act as good indicators of pollution generally.

Water quality control

Rani is employed as a scientist by the regional water company. Her job is to monitor the water quality by seeing which animals live there. She tells her story in the diagrams.

Back in the water company laboratory, the water is tested. One test measures **biological oxygen demand** (BOD). This involves measuring the amount of oxygen dissolved in a sample of water. The water is then kept in the dark at 20 °C for 5 days, and the amount of dissolved oxygen is measured again. The difference in oxygen content shows how much oxygen is being used up by bacteria in the water. The more bacteria there are, the more polluted is the water. So a high BOD value is a bad sign.

Further tests are done to measure the amount of solids suspended in the water and to discover what chemicals are dissolved in it. The water company has a duty to find the source of pollution of any chemical present in quantities over the legal limit.

All at sea

Pollution in the sea is difficult to deal with. Many countries legally dump materials into the sea. The enormous volume of seawater dilutes these chemicals and makes them harmless. However, there are concerns about how much longer places like the North Sea can remain a dumping ground without life in it being affected. Probably the most well-publicised incidents are oil tanker disasters. Spilled oil floats on the water surface and spreads out for miles.

1 I sample the water with a net and pour it into a shallow dish. The water animals soon show themselves. I identify and count them

2 This water is all right. It has a low level of pollution. That's because there are mayfly nymphs. Even purer water, with no pollution, would also have stonefly nymphs, caddis fly larvae and freshwater shrimps

3 The water louse, leech and alderfly larvae suggest that this water is moderately polluted. I suspect that there is run-off from the fields – probably fertiliser carrying nitrate and phosphate

4 Almost nothing in this water, but notice the bright red worms – bloodworms. They survive with very little oxygen dissolved in the water. That's a result of sewage pollution. I might find some rat-tailed maggots as well, but no fish or other animals that depend on dissolved oxygen. We need to act quickly here to stop more pollution

Sampling water quality

THE ENVIRONMENT

To minimise damage from oil several methods can be used:

- Use booms to trap spilled oil.
- Add detergents to break it down so it sinks to the sea bed.
- Add absorbent materials to mop it up.
- Burn it on the sea surface.

The victims of bad oil spills are usually sea birds, which dive into the sea to fish or rest on the sea. Oil sticks to their feathers, preventing them flying and reducing their insulating properties. Most oiled sea birds die from starvation and cold. Other animals, including fish, sea otters, and other mammals, can suffer as chemicals spread through the water. Short-term health problems and possible damage to vital organs have been reported by people involved in clearing oil spills.

Sea Empress spill claims 25 000 birds

Oil disasters at sea cause severe problems to wildlife

Cleaning up an oil spill

★ THINGS TO DO

1 Copy this table of water pollution into your notebook and complete it.

Level of pollution	Indicator animals
high	
moderate	
low	
none	

2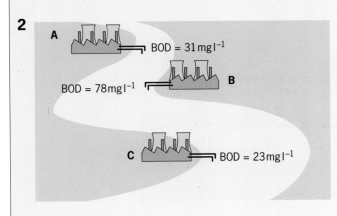

A BOD = 31 mg l^{-1}
BOD = 78 mg l^{-1} B
C BOD = 23 mg l^{-1}

The BOD of waste water released into a river from three different factories is shown on the map. The recommended limit for BOD in waste water is 25 mg l^{-1}.

a) Why is the water downstream of factory A likely to be polluted?
b) What would you look for in the water here?
c) Why could the water upstream from here be less polluted?
d) Identify other spots along the river where you would want to monitor water quality.
e) Why is a combination of biological signs and chemical tests used?
f) If possible, identify water animals taken from a nearby river or stream. Use the key in Topic 3.4 to help.

3 a) Contact your local council, farm, garage or industry to find out how they prevent pollution from harmful chemicals.
b) Contact your local water company and find out about water quality in your area.

3.20 In the air

> Air quality forecast for central Britain, Tuesday:
> Ozone 41 p.p.m. very good
> Nitrogen oxides 67 p.p.m. good
> Sulphur dioxide 60 p.p.m. good
> Benzene year average 2.11 p.p.m.
> (5 p.p.m. presents an exceedingly small health risk)
> 1,3-butadiene year average 0.32 p.p.m.
> (1 p.p.m. presents an exceedingly small health risk)

Air quality is monitored across Britain each day and the information is available on teletext. This can save lives. On one day in December 1991, about 160 people died in London from breathing and heart problems thought to have been caused by air pollution. In May 1995, many people suffered breathing difficulties when a photochemical smog settled over many places in Britain.

Pollutants include benzene, nitrogen oxides and 1,3-butadiene, caused by traffic exhausts. Low-level ozone is made in the air through reactions involving such gases, especially in sunny conditions and in heavy traffic at midday. On still days these are not carried away and become a health risk.

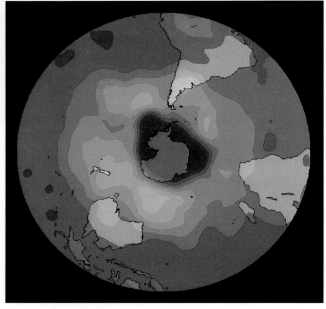

Satellite photo of ozone depletion over Antarctica

Benzene and 1,3-butadiene can cause cancer; ozone, sulphur dioxide and nitrogen oxides irritate the breathing system and make infections and asthma more likely.

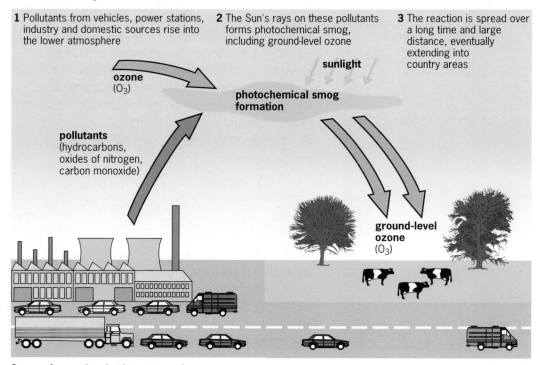

Ozone formation in the atmosphere

THE ENVIRONMENT

Problems caused by polluted air

In time the **greenhouse effect** may warm the planet enough to change weather patterns and raise sea levels. Its increase is due to excess carbon dioxide and methane in the atmosphere (see Topic 3.14).

Ozone depletion in the upper atmosphere allows more harmful ultraviolet rays from the Sun to reach Earth. These can cause skin cancers. **Chlorofluorocarbons** (CFCs) from aerosols, refrigerator coolants and aircraft also appear to destroy ozone.

Burning **fossil fuels** like oil and coal creates pollution in the form of tiny carbon particles ('particulates') in smoke, and harmful acidic gases including sulphur dioxide and nitrogen oxides. Vehicles also release acidic gases as exhaust fumes. These gases dissolve in water droplets in clouds, forming **acid rain**, which harms plants and animals.

The cost of reducing levels of pollution must be set against the cost of environmental damage. This relies on international co-operation, since the polluting gases may be blown hundreds or thousands of miles from where they were made.

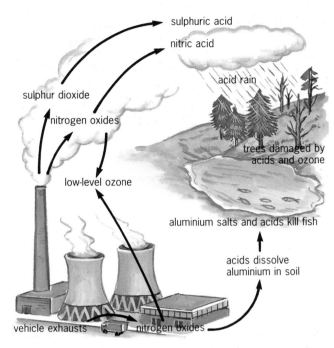

Acid rain formation

★ THINGS TO DO

1. Draw up a table to show where the following pollutants are made, and the problems each causes:
benzene, nitrogen oxides, sulphur dioxide, ozone and 1,3-butadiene.

2. The graph shows how ozone levels in a large city varied over 24 hours.

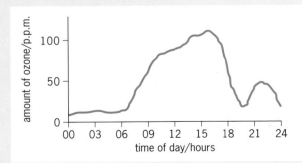

   a) Explain why the level of ozone in air varied over 24 hours.
   b) What advice would you give to someone with asthma living in that area?
   c) Keep a record of air quality over a week. If possible, plot the readings for different places on a map. Add some brief comments on why the air quality varied during the week.
   d) What steps could be taken to reduce the atmospheric pollution caused by cars?

3. Sulphur dioxide gas is released from food-sterilising tablets when water is added. It can be trapped in containers or plastic bags. But beware, it is a harmful irritant.
   a) How could you investigate the effect of sulphur dioxide on small plants (e.g. cress)?
   b) What changes might you measure?
   c) What could this tell you about acid rain?
   d) How else could you investigate its effect?
   e) Can you think of ways to measure the acidity of rain falling in your local area?

3.21 Working with ecosystems

Whaling begins again despite protests that species will become extinct

Farmers despair as rabbit population explosion slashes their crop yields

When ecosystems go out of balance there may be population explosions, or species may die out

We rely on ecosystems remaining in some sort of balance. The populations of different organisms within any community can change over time, but usually this takes place slowly, as new conditions may favour some plants or animals and ultimately lead to the evolution of new forms. On a shorter time-scale, shifts in population numbers can lead to wider distribution of favoured species, or conversely to extinction.

As the human population continues to rise rapidly, and lifestyles and expectations develop, further pressures are placed on the environment. Increased demands for energy and materials threaten more pollution and further damage to habitats.

Conservation

We are all responsible for the conservation of habitats and the living things in them. If future generations of humans are to enjoy the pleasure of looking at animals and plants in the environment, then we must take steps to protect it. **Conservation** means doing things to improve the survival of plants and animals and the place where they live. Many things can be done to preserve the rich variety of life forms that we see today (**biodiversity**).

An urban conservation area

THE ENVIRONMENT

Taking action over fish stocks

Stop the seas dying

If you stood at the bar of The Swordfish or The Dolphin in Newlyn 30 years ago you could tell by the singing when the boats were returning. Every dusk the sound of fishermen's voices, raised in song, drifted into the two pubs on the harbour front of the Cornish fishing village.

It was the sign of the end of a good day's fishing, and in those days – when there were, indeed, plenty of fish in the sea – every day was a good one.

(from *Independent on Sunday* 15 January 1995, with kind permission)

Overfishing has long threatened fish stocks around the world, but populations are now so low that fishing is no longer worthwhile in some areas. So few fish are caught that it does not pay the fishermen to go out (see graph). Many have sold their boats and lost their livelihood. Birds, seals, otters and other animals that eat fish also decline in numbers as their food becomes scarce.

Although fish populations in the seas around Britain are low, they will begin to increase if action is taken now. We could, for instance:

- limit the number of fish that can be caught by setting quotas,
- remove only fish above a certain size,
- not fish in areas where fish are breeding,
- reduce the amount of pollution released into the seas each day,
- pay fishermen compensation to take their boats out of the fishing fleet and find other work elsewhere.

This would allow fish to breed so that the numbers of newly born fish balance the number that die or are caught. Calculating sensible quotas is a difficult task, however, as no one really knows how many fish there are. Outright bans on fishing for certain species may be needed when stocks have dropped drastically. But ensuring that everyone agrees to the limits set cannot be guaranteed. Sadly, undersize fish are still caught and thrown away because they cannot be sold – a wasted resource and a threat to the species.

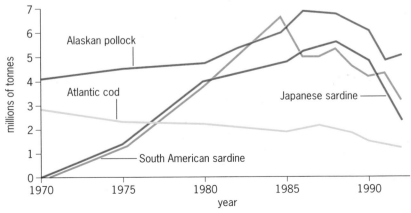

Changes in numbers of fish caught, 1970–1995 (from *Independent on Sunday* 15 January 1995, with kind permission)

WORKING WITH ECOSYSTEMS

Sustainable energy

Energy from fossil fuels cannot sustain our energy needs into the twenty-first century. The effects of pollution from burning them may also be too high a price to pay. The use of nuclear fuels would reduce the effects of atmospheric pollution produced by fossil fuels. But the risks of nuclear materials and wastes contaminating the environment and affecting health must be minimised and accepted.

Alternative energy sources (e.g. wind power, hydroelectricity, solar power) become better options as new technology makes them more efficient. 'Biodiesel', a fuel made from rapeseed oil, may offer a cost-effective alternative for cars and lorries. Other fuels, including alcohol and gas, may be the products of crops grown in the fields of the future. Cost will determine how successful they are. Increased costs of developing and using new energy sources must be set against the environmental costs of our present energy sources. (See *GCSE Science Double Award Physics* Topic 1.18.)

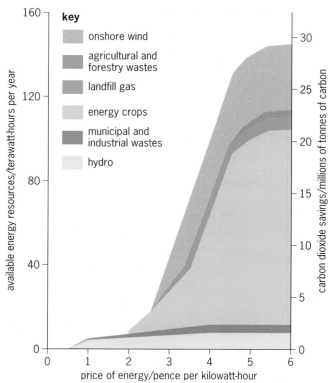

Carbon dioxide savings from non-fossil energy sources (from *New Scientist* 14 January 1995, with kind permission)

Major energy sources in future may include wind, water and plants such as rape

THE ENVIRONMENT

Sympathetic farming

In Britain about 80% of the land is used in agriculture. Farmers are increasingly under pressure to produce more food to feed an increasing population, and achieve high yields with minimum costs. This encourages the use of artificial fertilisers, pesticides and fungicides to ensure good, fast growth and to control less useful plants, insects and other animal pests.

Modern intensive farming methods are often unsympathetic to wildlife, however. Farmers frequently lack information on the environmental impact of changed farming practices. For instance, removing hedgerows to enlarge fields makes harvesting more efficient. But surveys suggest that wildlife populations can quickly decline as such habitats vanish (see Topic 3.11).

EU grants are paid to encourage farmers to leave wildlife refuges. Unfortunately their effectiveness may be cancelled out by other grants, such as money to drain wetlands or build dams, which destroys habitats.

EU grants may encourage habitat conservation, or discourage it: **a** site preserved from cultivation; **b** draining wetlands

All of these methods have an impact on the environment. Maintaining biodiversity requires farmers to appreciate how their work can affect plant and animal communities and plan to minimise any potentially harmful effects. It also means that we, as food consumers, have a responsibility to encourage good, responsible practices and to pay for them.

★ THINGS TO DO

1 Prepare a newspaper article describing international efforts to retain biodiversity. You might consider issues under headings such as: 'Time is running out to save the rain forests'; 'Osprey success threatened by egg stealers'; 'Supermarket chain to promote non-intensive farm foods'; 'Illegal traffic in wildlife on the up as elephants face extinction'; 'Conservation groups co-operate to protest at plan to build road through ancient woodland'.

2 Find out why larger supermarket chains are promoting food prepared by 'integrated crop management'.

3.22 Our world

By studying the environment we can understand how plants, animals and microbes interact to survive. The relationships within a community may be complex and dynamic, and it can be very difficult to predict how they may change, even when using complicated computer programs to model likely effects. It is often easier to explain what has happened after changes have taken place, and it is almost impossible to replace a community once a habitat has been destroyed.

An artificial world

Biosphere 2 is a large metal and glass dome, built in the desert in Arizona, USA, designed as an experiment to see if people could survive inside a self-contained artificial world. In 1993, 2 years after entering Biosphere 2, eight volunteers left the artificial world to breathe natural air again. They had survived despite problems in producing enough food and having to be given extra supplies of oxygen when levels dropped twice.

Further experiments could provide evidence of how to control ecosystems in the future or even lead to ways of living on planets like Mars!

- light gives energy for the plants to make food and oxygen, and allows the living things to be seen
- pump bubbles air into the water so oxygen dissolves for fish to breathe
- air bubbles carry small lumps of waste through charcoal, which absorbs the waste to clean the water
- some wastes settle at the bottom and are decayed by microbes
- food is added because there are no natural prey for the fish to eat
- any dead fish are removed quickly before they spread disease
- green algae growing on the glass tank have to be scraped off periodically because there are no herbivores to eat it

A tropical fish aquarium illustrates the principles of imbalance within ecosystems. This is an ecosystem that must be artificially maintained, otherwise it would stay out of balance and the organisms would die

Biosphere 2

THE ENVIRONMENT

Gaia

Our world is not artificial, but it may be far more complicated than was originally believed. One theory is that it is a superlarge 'organism' that can react to changes and survive. This 'Gaia hypothesis' suggests that environmental changes such as global warming and other effects of worldwide pollution will be cancelled out in time. As the world heats up, photosynthesising organisms will be able to use the extra carbon dioxide in the atmosphere. So the amount of carbon dioxide will eventually fall, resulting in less global warming. The world will cool down again. In essence, the planet will take care of itself, even if this takes a long time.

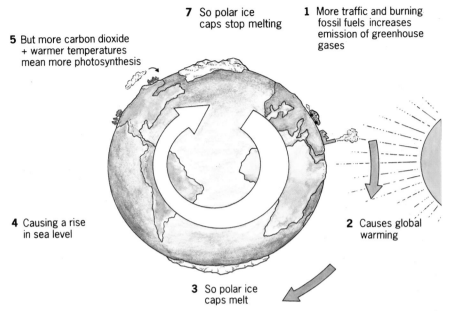

The 'Gaia hypothesis': the Earth as a 'superorganism'

★ THINGS TO DO

1 There are many good sources of information about the environment and groups that can help you learn more about our world. Find some useful addresses yourself. If you write for information, try to be specific about what you'd like to know and include a stamped addressed envelope.

2 Computer programs can be used to model the effect of changing factors within an ecosystem – natural or artificial.
   a) If possible, use a 'modelling program' to investigate how an ecosystem may change.
   b) Find out the advantages and disadvantages of using such techniques to predict environmental change.

3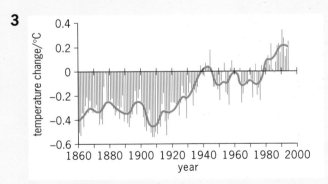

Rise in average global temperature since 1860 (From IPCC, with kind permission)

a) Look at the graph of average global temperature, and debate with friends the likelihood of the 'Gaia effect' changing the current rise in temperature.
b) Write an essay on the positive and negative influences of humans on the global environment.

Exam questions

1 A gardener used an old plastic dustbin to collect her garden waste. She wanted to convert the waste into compost.

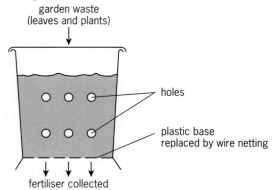

The contents are kept damp and when they start to decay some worms are added.
a) What organisms cause decay? (1)
b) Why is the garden waste kept damp? (1)
c) How do the holes in the side of the dustbin help the garden waste to decay? (1)
d) In the bin, the worms feed on the rotting leaves and waste and turn it into fertiliser. The fertiliser falls to the bottom of the bin so it can be collected. Why is it useful to turn these garden wastes into fertiliser? Give **two** reasons. (2)
(SEG, 1994)

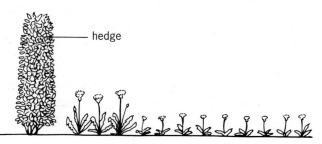

2 Two students were surveying dandelions in a field. They noticed that the dandelions by the hedge were taller than the others.
One student suggested that the differences in height could have been caused by the different conditions in the field.
a) i) Was he correct?
Give reasons for your answer. (2)
ii) Explain how you could test to see if his anwer was correct. (2)
b) The hedge was cut down and removed. What would happen to the height of the dandelions after some time? Explain your answer. (2)
(NEAB, 1994 (part))

3 Fig. 5 shows a food web for a pond.

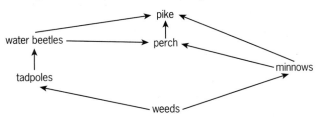

Fig. 5

a) Give an example of a primary consumer in this food web. (1)
b) [In the space below] Draw **one** food **chain** for this pond. (2)
c) A fishing club wants to increase the number of perch in the pond. Using the information in Fig. 5, suggest how they could do this. (1)
(ULEAC, 1995)

4 In 1950, near Long Island, USA, a group of scientists studied the breeding colonies of the Osprey. Ospreys are birds that only eat live fish. The food chain below shows a source of the Ospreys' food.

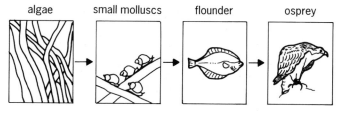

a) From this food chain, write down an example of:
i) a producer; (1)
ii) a primary consumer. (1)
iii) It is possible to construct a pyramid of numbers for this food chain. How would the number of molluscs compare with the number of Osprey? (1)
b) The graph shows how the number of breeding pairs of Osprey in a colony near Long Island changed between 1950 and 1980.

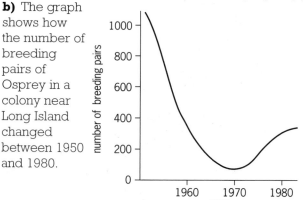

i) By 1965 local people believed the sea near Long Island was polluted. Using the previous information what evidence was there to suggest that the sea was polluted? (1)

ii) It was found that part of the "pollution" of the sea was caused by a pesticide called DDT. This was being used to kill mosquitoes which lived in the area. From the graph, find when the use of DDT was stopped. (1)

iii) It was found that the concentration of DDT in an Osprey was much greater than that in the algae. Explain why. (1)
(SEG, 1994)

5 The diagrams show maize plants grown from seeds sown at different distances from each other.

a) Write down **two** differences you can see between plants A and B. (2)

b) The differences are caused by competition between the maize plants.
The maize plants are competing for **light**. The maize plants are also competing for .... and .... (2)
(NEAB, 1995)

6 a) The diagram below shows five organisms in their aquatic environment.

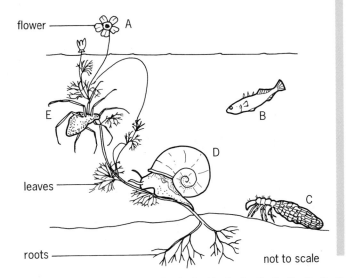

Use the key below to place the organisms A, B, C, D and E in their correct group.

KEY
1 (a) Has fins   Stickleback (fish)
  (b) Does not have fins   go to 2
2 (a) Has legs   go to 3
  (b) Does not have legs   go to 4
3 (a) Has six legs   Caddis fly larva (insect)
  (b) Has eight legs   Water spider
4 (a) Has a hard shell   Ramshorn snail
  (b) Does not have a hard shell   Water crowfoot (plant)

In the table below, use a tick to show each step you have used. Organism D has been done for you.

Organism	1a	1b	2a	2b	3a	3b	4a	4b	Group
A									
B									
C									
D		✓		✓			✓		Ramshorn snail
E									

(4)

b) Outline the functions of the organs labelled on organism A.
One function of the **leaves**.
Two functions of the **roots**. (3)
(ULEAC, 1994)

7 A farm is an ecosystem managed by a farmer. The crops grown at Green Mile Farm are barley, sugar beet and beans. Some organisms are pests which reduce the yield of the crop plants. Some pests, for example aphids, feed on crop plants and can pass viruses on to them.
The farmer at Green Mile Farm wanted to make as much profit as possible. Table 1 below shows details of local pests. Table 2 shows agricultural chemicals which farmers can use.

Use the information above to answer the following questions.
a) Which agricultural chemical would the farmer use to increase crop yield if there were no pests in the area? How would this help to achieve a greater yield? (2)
b) Suggest **three** ways in which the farmer could prevent the fungus 'Take all' from harming his barley crop. (3)
c) Table 1 states that couch grass competes with crops for resources.
i) How does competition lead to reduced crop yield? (2)

EXAM QUESTIONS

Table 1

Organism which reduces crop yield	How it over-winters	Crop affected	Effect
'Take all' (fungus)	Lives on unploughed stubble and couch grass	Barley	Uses up plant's carbohydrates
Black bean aphid (insect)	Lives on weeds in hedgerows	Beans	Attacks the growing tips of bean plants
Couch grass (weed)	As seeds and underground stems	All crops	Competes with crops for resources
'Yellows' (virus)	In stores of sugar beet and mangels	Sugar beet	Attacks leaf tissues which turn yellow
Beet aphid (insect)	Lives in beet stores and on weeds	Sugar beet	Penetrates tissues passing on viruses

Table 2

Agricultural chemical
Fertiliser
Insecticide
Fungicide
Herbicide

ii) What could the farmer use to get rid of the couch grass? (1)

d) i) When attacked by the 'yellows' virus, sugar beet leaves turn yellow. How will this reduce crop yield? (2)

ii) Viruses cannot move independently. How does the 'yellows' virus enter the leaf tissues of the sugar beet? (2)

iii) How could the farmer prevent the 'yellows' virus from attacking the sugar beet crop? (2)

e) Suggest **one** environmental factor which should concern the farmer who uses agricultural chemicals.
(1)
(ULEAC, 1995)

8 At present, fish farming in Scotland produces 70 000 tonnes of salmon each year.
Eggs are hatched under laboratory conditions and the young fish are later transferred to large net cages in a freshwater or sea loch or lake. A cage can hold 30 000 to 40 000 fish.
Here they are provided with a high energy, high protein feed until they have grown to a marketable size.

Antibiotics are also supplied in the feed.

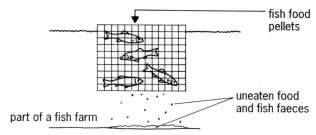

part of a fish farm

a) What are the advantages of fish farming compared with traditional fishing. (4)

b) One problem with fish farming is that the crowding together of large numbers of fish results in their being affected by parasites called sea lice. These do not kill the fish but cause severe skin wounds and slow down the growth of the fish.
The sea lice are killed by treating the affected cages with a pesticide.

i) Explain how parasites and their control reduce the profits of the fish farmer. (2)

ii) An alternative treatment could be to use fish called sea wrasse to pick off the lice from the skin of the salmon.
Give **two** advantages of this method compared with using pesticides. (2)

c) One possible development will be to use net cages that can be towed behind a ship to a position in the sea or ocean where plankton (microscopic plants) is abundant. These ocean cages can be towed to new sites from time to time.
What benefits will this system have compared with the present method of fish farming? (2)
(ULEAC, 1994)

REPRODUCTION, INHERITANCE AND EVOLUTION

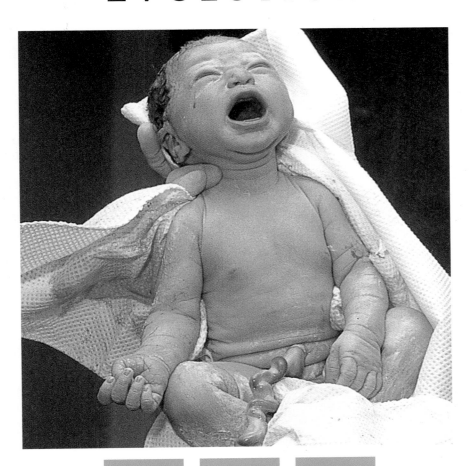

4.1 Sorting organisms out

When you walk into the supermarket you know where things are because the different types of product have been grouped together. That makes it easier for you to find the things you need. Meats are grouped together. Frozen foods are found somewhere else. The fruit and vegetables will be in a different place again. Living things are classified (sorted) in a similar way.

Classification

'Classification' is the name given to the process of sorting things into groups. Each group will have certain similarities (things that they have in common). Most living things fall into five main groups – microbes (bacteria, viruses), single-celled creatures, fungi, plants (the **Plant Kingdom**) and animals (the **Animal Kingdom**). But even within each group the living things, although similar in many ways, may appear quite different.

Members of the Animal Kingdom: **a** a sea anemone; **b** a snail; **c** an eagle; **d** a human

The Animal Kingdom

It is impossible to show in a book the wide variety of animals that exist on earth – there are millions of them. We can, however, get some idea of this by comparing a few.

You can see some animals in the photographs. Whilst the sea anemone looks more like a plant, others are quite clearly animals. To help us classify them in some sensible way, we put them in smaller groups, by looking at their *characteristic features*.

Firstly, all animals either have a backbone or do not. So we can form two smaller sub-groups of 'animals without backbones' (**invertebrates**) and 'animals with backbones' (**vertebrates**). Even those within the same group, however, are quite different. So these two groups need to be broken down further into even smaller groups. This is shown in the diagram.

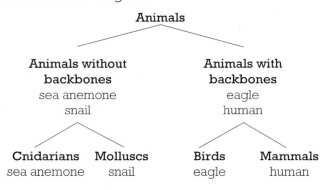

Each small sub-group may in turn contain hundreds of different types of animal. The group of birds, for example, will also contain starlings, sparrows, hawks, cormorants and many other birds. All of them have these things in common:

- skin with feathers,
- eggs with hard shells,
- wings (or the remnants of wings), which may help them fly (although there are exceptions, such as the ostrich),
- a beak for feeding,
- lungs for breathing,
- warm blood.

REPRODUCTION, INHERITANCE AND EVOLUTION

The Plant Kingdom

The living things that make up the Plant Kingdom can be grouped in a similar way, depending on their features.

The plants in the photographs have quite different characteristics, including their appearance. For example, the type of tree shown (ginkgo) and the fern do not flower. The tree produces seeds in 'cones'. The fern produces reproductive spores on the underside of the leaves; the leaves die back each year. The flowering plant, however, will die back each year after producing seeds, which will germinate the following year. These, and other features, allow us to form sub-groups of plants as follows:

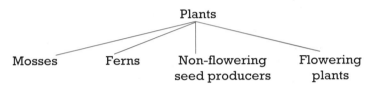

Each of these groups will contain many others. The group 'flowering plants' will, for example, include a wide variety (from trees to bulbous plants) of plants that produce flowers. And these can in turn be sub-divided; for example, bulbous plants include crocuses, daffodils and tulips.

Members of the Plant Kingdom:
a a fern; **b** a garden perennial; **c** a ginkgo tree

★ THINGS TO DO

1. Make a classification system for a supermarket based on the types of produce they sell.

2. **a)** Collect photographs of as many different animals and plants as you can. Sort them into groups based on similarities and differences, not only in appearance, but also in lifestyle – where they live, what they eat, how they reproduce and so on. How many different groups can you make?
   b) Make a key (like the one in Topic 3.4) to help classify the organisms in the photographs.

3. The system of giving all living things a common name (in Latin) was suggested by a Swedish scientist called Carolus Linnaeus in 1735. The table shows different names for the same plant or animal.

   a) What are the advantages of a common system of names?
   b) Use some detective work to find the missing names.
   c) Add some names of other living things and challenge a friend to complete the table!

Latin	English	American	French	German
Gavia stellata	red-throated diver	red-throated loon	plongeon catmarin	Sterntaucher
Fraxinus excelsior	European ash		frêne	
Digitalis purpurea	foxglove			
Hirundo rustica			hirondelle de cheminee	Rauchschwalbe

4.2 It's all changed

Dead as a dodo

We've all heard the term 'as dead as a dodo'. The dodo was a bird that lived for millions of years in the area around Mauritius on the east coast of Africa. Like the ostrich, it could not fly.

Today the dodo does not exist as the species has died out (it has become **extinct**). There are many possible reasons for this:

- It may have lost the ability to protect itself from predators.
- Humans caught it for meat; they may eventually have killed every one.
- Animals introduced by humans may have preyed on its eggs and the young chicks.
- As humans developed the land, they may have destroyed the dodo's habitat.

The fate of the dodo is similar to that of many other animals and plants, such as the dinosaurs, that also became extinct. Their fate was determined by changes on Earth; some of these were natural (as in the case of dinosaurs), but others were caused by people.

The dodo

The changing Earth

Earth was formed 4600 million years ago (see *GCSE Science Double Award Chemistry*, Topic 2.3). At that time no living thing could have survived. Slowly, however, Earth changed to provide the things that living things need – water, oxygen, food and non-extreme temperatures. As it changed, so did living things. Some adapted as the conditions changed; others, for one reason or another, died out. Some of these changes appear quite remarkable. The London we see today was once a tropical swamp, inhabited by alligators, tigers and mammoths. Modern-day desert regions were once fertile lands where animals and plants flourished. There are lots of different ways in which evidence can be gathered to show that these changes have taken place.

These fossilised remains of dinosaurs were found in rocks in the Aude, France

Evidence from fossils

Fossils of animals and plants that lived millions of years ago give us many clues. Most are found in sedimentary rocks. The fossil in the photograph was found in limestone. This rock forms from the remains of sea creatures, so

These were drawn by people who lived on Earth thousands of years ago. Their paintings show the animals that once existed in the area

REPRODUCTION, INHERITANCE AND EVOLUTION

Fossils in limestone rock

In the Grand Canyon, fossils found in the lower layers of rock lived on Earth long before those found in the higher layers

the animal must have lived under water. Fossils of sea creatures can be found even high up on the slopes of Mount Everest, so this site must also, at some time, have been under water, and later on been pushed high above sea level by movements of the Earth's surface.

Throughout time, layer upon layer of rock formed from dust and sediments settling on the land surface and the sea bed. The remains of animals were trapped in these layers. By studying fossils found in different layers of rock we can see how they have changed and adapted through time. The depth of the layers tells us how old the fossils are.

Tests on the bones of skeletons (see *GCSE Science Double Award Physics*, Topic 4.16) can tell us how old they are. The size of the bones, the type of teeth and many other features tell us what the animal ate and what the environment was like at that time.

Ice taken from below the North and South poles, as in the photograph, contains pollen, which tells us what types of plants existed when the ice was formed. The deeper the ice, the longer ago it was formed. It may be up to 100 000 years old. Dust particles and gases trapped in the ice tell us which chemicals were in the air, and how much of each there was.

Extracting ice samples: the thickness of the ice layers tells us what the climate was like at that time

The remains of this person were found preserved in a peat bog in East Anglia (a lack of oxygen prevented decay). Even the food remaining in his stomach could be identified, providing information about his diet. Discoveries of ancient human bones and remains show how very different humans were in the past

155

IT'S ALL CHANGED

This invertebrate is one of the oldest ever found; it lived in the seas 500 million years ago

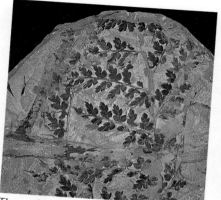

These fossils of plants suggest that they were from the time when the first creatures lived on land; they appeared 300 million years ago

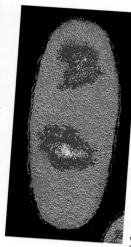

The first bacteria and algae appeared some 3000 million years ago

Dinosaurs lived around 190 million years ago; this period (the Jurassic) was called the 'age of the reptiles'

A time clock

We know that the living things that we see on Earth today, and those that are now extinct, have formed over 3 billion years from very simple organisms.

The world about 100 million years ago was covered with vegetation, and animals roamed freely; flowering plants also appeared at about the same time

The first mammals appeared about 70 million years ago; many, like the woolly mammoth, are now extinct

The dating of the bones from the first humanoids (human-like mammals) suggests that they appeared around 1.5 million years ago. Flowering plants and other mammals also flourished then

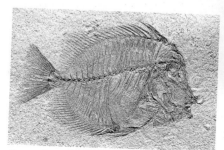

Bony fish like this lived in the seas around 55 million years ago

This rock is thought to be the oldest known, 4200 million years old in fact, formed just after the Earth was created 4600 million years ago. At that time much of Earth was covered by volcanoes, and the atmosphere was full of poisonous gases; the first, simple living things could probably survive only in the seas

REPRODUCTION, INHERITANCE AND EVOLUTION

How were fossils formed?

When an animal dies the softer parts of the body normally decay quickly (see Topic 3.12) because of the action of oxygen, bacteria and enzymes. The harder parts, such as bones and teeth, are left behind, although even these will decay after a long enough time.

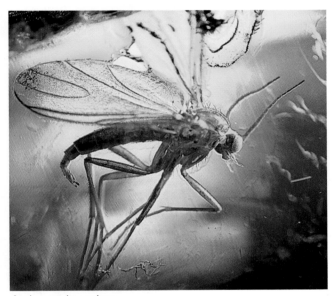

An insect in amber

Sometimes, minerals in the surrounding rock may penetrate the pores in the bones, turning them to stone; many tree fossils have been formed in this way.

Sometimes, a dead organism has been covered by mud; the body then acts as a mould for the mud, which forms around it and hardens. Over time the creature's remains decay, leaving a space in the mud, which gradually turns to stone. When the rock is broken, the hollow mould still shows the shape of the original animal, including bones and even feathers.

Sometimes this mould becomes filled with chemicals from the surrounding rocks. These harden, forming a cast in the shape of the original mould. The perfectly preserved remains of early insects have been found in amber (the fossilised sap of pine trees). Other remains have been preserved in peat, and tar oils, which contain natural preservatives. Woolly mammoths and sabre-toothed tiger remains have been found in ice and tar pits in Russia; other remains appear as glaciers melt.

★ THINGS TO DO

1 a) Sort out the evidence illustrated in the 'Time clock' section into a chart showing the main events and the time at which they occurred. You can start with:

4600 million years ago: the Earth was formed from dust and gas

Try to make your chart attractive by adding some illustrations of the major events.
b) Which came first, plants or animals? Why do you think they appeared in that order?
c) Why do you think the first animals would have flourished on Earth?
d) What would have happened as the number of animals increased?
e) What would animals and plants have needed to survive? What does this tell you about the Earth at the time?
f) When the first humans appeared on Earth what would they have found? What features would have helped them to spread as quickly as they did?
g) What will be found in millions of years' time that tells people about the world as it is now? Where will it be found?

2 Many people believe that some animals such as whales, tigers and gorillas may become extinct over the next 50 years.
a) Make a list of reasons why animals such as these, and plants, may become extinct.
b) How can 'seed banks' such as the collection at Kew Gardens, London, prevent plant species from becoming extinct?

3 Try to find out reasons suggested for why the dinosaurs became extinct.

4.3 Selection and evolution

Darwin's fantastic idea

In 1835, Charles Darwin, a naturalist sailing on a ship called the 'Beagle', visited a small group of islands (the Galapagos Islands) in the Pacific Ocean. He studied the wildlife on each, and found several different species of finches, a type of bird. He noticed that the birds on different islands had many similar features to each other, and to the common species of finch on mainland South America some 1000 miles away. But he also saw some striking differences.

Darwin was fascinated with similarities and differences among these birds, and other animals he studied on his 3-year journey. He proposed a fantastic idea to explain their origin. In contrast to the teachings of religion at the time, which were of great influence, he said that each animal had changed, or **evolved**, from animals that had lived before. He also suggested that the finches had evolved from one species of finch, a 'common ancestor'. Here is a summary of his ideas:

- Reproduction produces a variety of different offspring (i.e. children) (see Topic 4.5).
- There is competition for food, breeding territory, etc. between members of a species.
- Individuals with characteristics that give them a competitive advantage over others of their species are more likely to survive; others die from predation and disease.
- By breeding with similar surviving members, 'successful' characteristics are passed on.

So these characteristics are effectively 'selected' by the environment. This process is termed **natural selection**.

Darwin realised that his ideas were so new, and offensive to many people, that he delayed publishing them until 1858. His fears of criticism from religious groups were justified: some faiths refuse to accept his views even today. (Opposition to his ideas is particularly strong in parts of the United States.) But modern-day scientific understanding of the origin of all living things and their continued evolution owes a great deal to Darwin's hypothesis. In addition, we now know (which Darwin didn't) how information about the characteristics of organisms is passed on by genes (see Topic 4.4).

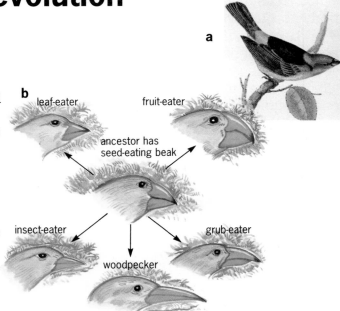

a One of Darwin's finches. **b** The first species of finch on the Galapagos Islands was thought to have a beak adapted to eating seeds. Individuals of this species then flew to other islands. Those with a slightly different beak, who could eat new foods, had a competitive edge. Up to 16 species now exist

Changing all the time

Evidence from fossils (see Topic 4.2) shows that living things have been changing for millions of years. Natural selection can help us to understand both changes that take place slowly and those that happen suddenly.

Evidence for selection over a much shorter period of time, from 1848 onwards, comes from looking at populations of the peppered moth, *Biston betularia*. There are two forms, light and dark coloured. The colouring provides camouflage against the bark of trees, as shown in the photographs. Light moths are well camouflaged against lichen-covered bark, but dark moths stand out against it, and are more readily eaten. However, when soot from coal fires increasingly turned the trunks of trees black in industrial areas up to the 1950s the dark forms were better camouflaged. With the passing of 'clean air' acts at this time, this situation changed again. The map shows how the populations of light and dark moths were distributed earlier this century.

REPRODUCTION, INHERITANCE AND EVOLUTION

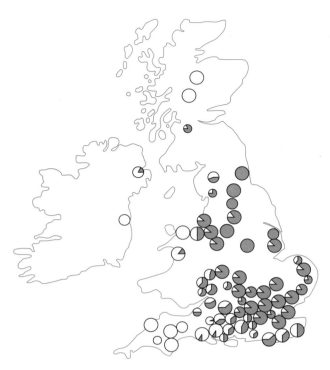

Light and dark moths against light and dark trees

Biston betularia distribution in the 1950s (from B Kettlewell, University of Oxford, with kind permission)

★ THINGS TO DO

1 a) Explain how the population of dark moths is related to smoke pollution.
b) What changes would you predict in an area after this pollution has been reduced?
c) How could you monitor the population of light and dark moths in an area?
d) How could you investigate the preference of birds for eating different-coloured foods (e.g. moths)? Could you extend this to test the influence of background colours?

2 a) The selection process can be modelled using a computer program. If possible, use software to investigate how selection works.
b) An alternative method is to use a patterned tablecloth (or equivalent) to represent the background and coloured counters as prey animals (see Topics 3.9, 3.10). By acting as a predatory animal, picking off prey according to their colour, you can 'simulate' selection. After a certain time interval you can allow for breeding by adding new counters.
i) Try to predict the effect of 'eating' prey of one colour, or all prey except one colour. See what happens when you vary the 'rules' of selection (such as light or dark background advantage), or the breeding rate. Think of other conditions that you might alter.
ii) Find out how long it takes for one type of prey to increase, or another to become extinct.
iii) Look for patterns to explain how selection operates.

3 Although Darwin's ideas were put forward more than a century ago, they are still treated as only a 'working hypothesis' by many scientists, as they do not account for a number of important problems. Try to read some modern books on evolution and selection by people with opposite viewpoints, such as Richard Dawkins (e.g. *The Selfish Gene*, *The Blind Watchmaker*), Niles Eldridge (*Reinventing Darwin*), Stephen Jay Gould (e.g. *Wonderful Life*, *The Panda's Thumb*) or Brian Goodwin (*How the Leopard Changed its Spots*).

Answer (or discuss as a group) the following questions.
a) What major problems in evolution do Darwin's ideas not explain?
b) Whose ideas do you think provide the best explanation for evolution and why?

4.4 We're all different

Each species of living thing can be recognised by their features (see Topic 4.1). Some are clearly seen; you recognise a dog and a cat because they are quite different in appearance (although there are many features of their lifestyle that are common to both). Others are not so apparent; the cat and dog, for example, are both **mammals** – they are warm blooded and give birth to live young.

These 'cat features' have developed over millions of years, and are passed on from generation to generation – they are **inherited**

The flower colour of hydrangeas depends on **environmental** influences (the soil in which they grow): in alkaline soil they are pink, while in acidic soil they are blue

Variation

Differences between different species of plants or animals are usually obvious. They are less obvious between members of the same species, but individuals are never *exactly* the same. For example, every person on earth is different to every other person. The differences may be very small, such as those between 'identical' twins (see Topic 4.14), or great, such as those between a tall female and a small male. The differences between individuals are called 'variations'.

Some features show an either/or variation; for example, you are either female or male, and you can either roll your tongue or not (see Topic 4.15). This is called **discontinuous variation**.

Some features can show a range of variation, for example, your height or weight. This is called **continuous variation**.

Discontinuous variation is the result of genes being inherited from parents. Continuous variation is often caused by both genetic and environmental influences. For example, your weight depends on inherited features of your body such as bone size and density, but it is also affected by lifestyle and food availability.

Information in code

The bar codes on goods in a shop contain information about what they are, where and when they were made, and their current prices. Those on foods may also contain information about the sell-by date. Although we cannot see it, this information is coded and stored in the magnetic stripes.

This group of children show both discontinuous variation (they are either male or female) and continuous variation (e.g. in their heights and weights)

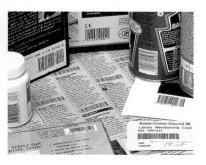

Information about goods is stored in a bar code

All living things also carry coded information. In the nucleus of every cell of the human body, there are 46 long thin strands called **chromosomes**, made up of a substance called **DNA (deoxyribonucleic acid)** plus protein (see Topic 4.6). Each chromosome is similar to the bar code described above – that is, it contains information. The chromosome is organised into sub-units called **genes**, many of which (either singly, or in combination) carry specific information about a characteristic of the organism (for example, hair colour). Other genes control the body's production systems (e.g. protein manufacture), or even switch other genes on and off. Some seem to have no obvious effect.

★ THINGS TO DO

1 A class made a note of all the differences between them.

Sex	Height/cm	Tongue roller?	Shoe size	Weight/kg
m	172	y	10	58
m	162	n	7	45
m	174	y	10	57
f	148	y	2	44
f	176	y	8	55
f	166	n	7	58
m	162	y	5	45
f	171	n	6	58
m	163	y	7	49
m	153	y	6	45
f	167	y	5	50
f	171	y	6	55
m	177	y	10	65
m	168	y	10	66
f	150	y	6	49
f	154	y	8	55
f	163	n	7	60
m	161	y	7	46
m	157	n	6	44
m	158	n	5	42
m	162	y	7	61
f	169	y	6	52
f	168	n	6	55
f	158	y	5	51

a) Which features show discontinuous variation?
b) How does an individual get these features?
c) Copy and complete this frequency table for the data on height, then draw a bar chart to show the variation in height. Do this first for boys, then for girls.

Height range/cm	Number of people
145–9	
150–4	
155–9	

i) What is the most common height range?
ii) What does the graph tell you about very small and very tall people?
iii) How would this type of data be useful to clothes manufacturers?

2 a) Measure the longest side of each bean seed in a packet. Put your results in a table like this:

Length of seeds/mm	Number of seeds	
	Tally	Total
0–5	III	3
6–10		
11–15		

b) Use the data to draw a bar chart.
c) What can you say about the variation in bean size?
d) Measure and record other differences between the seeds.

4.5 Reproduction

Asexual reproduction

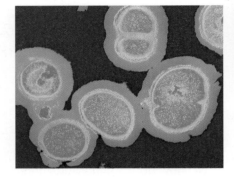

Bacteria reproduce by simply dividing into two. As they do so each new bacterial cell receives a full set of genetic material. Other microbes and simple organisms can also reproduce like this. For example, bacteria reproduce very quickly in warm milk, causing it to become spoiled – or 'go off'.

The type of reproduction shown in the photograph is called **asexual reproduction**, because only one parent is needed. As well as microbes, a number of plants can reproduce asexually, so this type of reproduction is sometimes called **vegetative propagation** (see Topic 4.7). Because no other plant has been involved, the new plants will contain genetic information from only a single parent. They will all be exactly the same and identical to the parent – they are called **clones** of the parent (see Topic 4.8). Commercial plant growers may use this to obtain large numbers of selected varieties of plants that are resistant to disease and crop well.

Sexual reproduction

How often have you heard the saying 'You look just like your mum (or your dad)'? You look and behave like them in many ways because many of their characteristics were transferred to you in the form of genetic information in the chromosomes. But many of your characteristics will also be yours alone and different to either of your parents. This is because you are the product of **sexual reproduction** – a process that needs two parents and increases variation. Most plants and animals rely on sexual reproduction to produce offspring (see Topics 4.9 and 4.10). So new generations are similar, but not identical, to earlier ones.

Similarities between a parent and child

Which is best?

Many plants, and some animals, reproduce both sexually and asexually. Each has its uses.

Asexual reproduction produces identical offspring because the offspring receive the same information (in genes on chromosomes) from one parent. In the simplest form, as seen in the bacterium above, a cell just divides, by a process called **mitosis**, and two new cells are made. Each 'daughter' cell is given a full set of chromosomes, identical to that of the dividing 'mother' cell, as shown in the following diagram.

Sexual reproduction usually produces variation in offspring. Special sex cells, called **gametes**, are produced in the male and female sexual organs by a type of division called **meiosis**. This halves the number of chromosomes received by each sex cell (so is also called 'reduction division'). This reduction is essential, so that when the sex cells (sperm and egg) later join up in **fertilisation**, the normal chromosome number is restored. For instance, humans have 46 chromosomes in each body cell (this is termed the **diploid number**); meiosis reduces this number to 23 chromosomes in each sex cell (termed the **haploid number**).

REPRODUCTION, INHERITANCE AND EVOLUTION

This number varies from species to species, but is always constant for a particular species, except in special circumstances (e.g. genetic disorders such as Down's syndrome – see Topic 4.6).

After fertilisation, further repeated cell division by mitosis then produces the millions of cells in the body. This ensures that each new cell in the body contains all the 46 chromosomes that came together in the fertilised egg (called the **zygote**). In other words, each cell has a full set of information inherited from both parents equally. As a result the new individual is unique – that's why there is not another person like you in the whole world. Other people are programmed with different information.

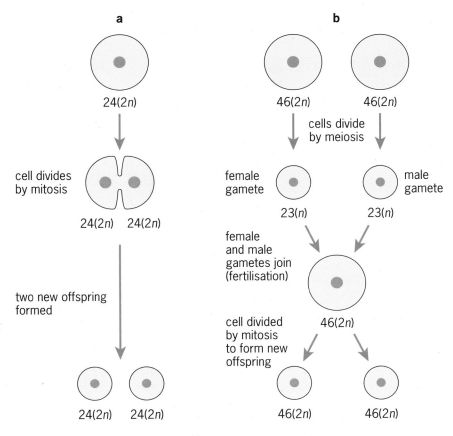

a Asexual reproduction; **b** sexual reproduction

★ THINGS TO DO

1 You can watch organisms such as yeast reproducing asexually using a microscope. Yeast cells 'bud' as they multiply. Follow these instructions:
a) Use a dropping pipette to put a drop of yeast suspension on to a glass slide.
b) Add a drop of stain such as methylene blue.
c) Place a cover slip over the drop on the slide.
d) Examine your slide under a microscope. (See 'Using a microscope' in the Teachers' Guide.) Draw anything you see that suggests the yeast is multiplying.

2 Greg had a plant with red flowers. He planted seeds from this plant. They grew into new plants with red flowers. He collected seeds from these and planted them the following year. The seeds he planted in the flower bed grew into big, strong plants. Some had red flowers, others white. He had dropped some seeds on the gravel path during planting. These grew into small weak plants, some of which had red flowers whilst others had white. The leaves of all the plants were the same shape, but different sizes.
a) Why did all the plants have the same shaped leaves?
b) Why were some flowers white whilst others were red?
c) Why did some plants have small leaves whilst others had larger leaves?
d) What do you think would have happened if the seeds that fell on the path had fallen on the flower bed? Explain your answer.

3 If possible, observe chromosomes in a dividing cell under a microscope or on a CD-ROM. Draw what you see and count the number of chromosomes.

4.6 Two kinds of division

Mitosis

When cells are not dividing, the chromosomes containing the DNA are in a long, thin, stretched-out form, and are invisible. Just before cell division begins the amount of DNA in each chromosome doubles as each chromosome copies itself, a process that is called **replication**.

This child has Down's syndrome, a condition where the body cells contain an extra chromosome (making 47 in all). One cause of it is chromosomes not being shared out equally when sex cells are made. Although someone with Down's syndrome can lead a happy life, they may suffer from a variety of physical and mental problems. Tests can now show whether a fetus has this condition, and abortion may sometimes be considered

Chromosomes come in pairs (they are called **homologous** pairs); the 46 chromosomes in human cells form 23 pairs. One of these pairs (the sex chromosomes) determines whether you are boy or girl (see Topic 4.14). If you look at a particular pair of chromosomes in detail then you will see that at the same position on each there is a gene that carries information about one particular characteristic. The alternative versions of a particular gene are called **alleles**. It is the combined effect of alleles that determines what an organism's characteristics will be (see Topic 4.15).

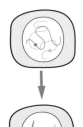

1 The nucleus of each body cell has 23 pairs of chromosomes, making 46 in total

2 Chromosomes shorten and thicken

3 Two copies (chromatids) of each chromosome become visible as they thicken and contract

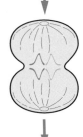

4 Chromosome pairs line up in the middle of each cell, with partners in each chromosome pair next to each other

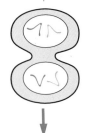

5 When a cell divides into two a copy of each chromosome passes into each new cell so both 'daughter cells' contain 46 chromosomes – each cell has identical chromosomes

6 The two daughter cells separate

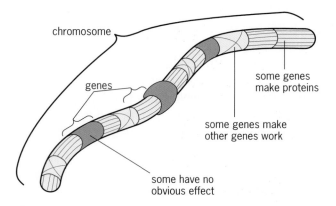

Genes on a chromosome

The stages of mitosis

REPRODUCTION, INHERITANCE AND EVOLUTION

At the start of cell division the chromosomes become visible as they shorten and thicken. They can be seen to be composed of a pair of parallel strands; each single strand is called a **chromatid**. The sequence of events shown in the previous diagrams then takes place; it was worked out by observing cells dividing. (It shows only one pair of chromosomes, for clarity.)

Meiosis

The process of meiosis also contains a number of stages, of which the initial shortening and thickening of the chromosome is as in mitosis. However, two different divisions then take place, so that from each cell four daughter cells are formed, but each has only half the original chromosome number. This process is essentially the same in plant and animal cells. (The diagram illustrates meiosis in a cell with one pair of chromosomes.)

Introducing variation

Chromosomes from each parent are separated into different sex cells during meiosis. They are recombined when sex cells join up in fertilisation. This creates new combinations of features in the offspring. Variation is increased further if crossing over between chromosomes takes place during meiosis. The diagram shows how this can happen.

1 Chromosomes become visible in the 'mother' cell; a network of fibres holds on to the chromosomes and helps them to move around the cell

2 The two copies (chromatids) of each chromosome become visible as they thicken and contract

3 Each pair of chromosomes lines up in the middle of the cell, with partners in each chromosome pair next to each other (these may cross over – see following diagram)

4 Fibres pull on the chromosomes, forcing each chromosome in a pair to opposite ends of the cell

5 Two cells are formed, although they may not separate completely

6 The chromatids on each chromosome separate, and four new 'daughter' cells are made. The chromatids will develop into chromosomes. Each cell has half the number of chromosomes of the 'mother' cell

The stages of meiosis

3 Two chromatids from a pair of chromosomes overlap; there is a chance that the chromatids will break, and join up again — if the 'wrong' ends join up (recombine) then two new chromatids are produced

4–6 The stages of meiosis continue, producing four gametes. But two gametes now have different chromosomes to the parents (they are **recombinant chromosomes**)

cell with parental chromosome   cell with parental chromosome

cell with recombinant chromosome

Crossing over in meiosis (stages 3 to 6)

TWO KINDS OF DIVISION

DNA

Deoxyribonucleic acid (DNA) is a large, complex molecule, consisting of two long strands, coiled together to form a 'double helix'. Each strand is a chain of small units called **nucleotides**. A nucleotide is made up of three chemical groups: phosphate, sugar (deoxyribose) and a base. There are four types of base called adenine (**A**), guanine (**G**), cytosine (**C**) and thymine (**T**).

Strong chemical bonds hold these units together as a long chain. Weaker hydrogen bonds between complementary bases (AT, CG) on opposite strands hold DNA together as a double helix.

The sequence of bases along a strand of DNA can be very variable. It is this sequence that carries information, coded as 'triplets' of bases. Each triplet, a **codon**, may code for a specific amino acid (which will later be added to other amino acids to make a protein molecule). Some codons carry other instructions, like full stops to end messages, or starting points. So a gene consists of a part of a DNA molecule made of many codons.

DNA replication

Before cells divide each chromosome doubles up. This requires each molecule of DNA making up the chromosome to produce an exact copy of itself. The copying process is called **replication**. It starts when a DNA double helix is unwound by a special enzyme, in a similar way to a zip being undone. The bases that were loosely bonded together in the double helix become exposed and now act as templates. A further enzyme, DNA polymerase, adds a complementary sequence of bases to each template to build up a new chain. Each new chain and template then wind up together as a new double helix, so two new identical DNA molecules are formed. The figure below shows how this happens.

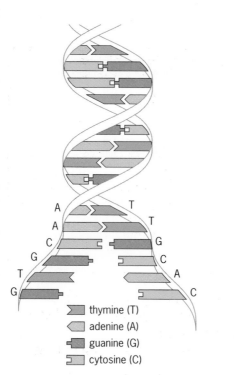

How the DNA molecule is made up

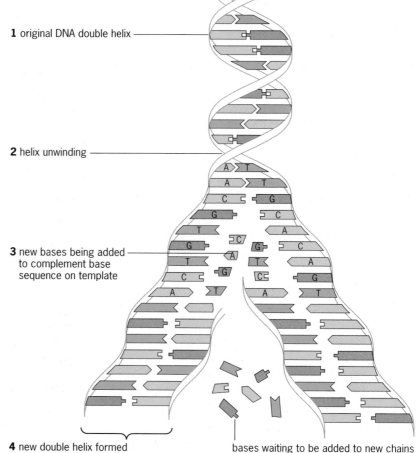

DNA replication

REPRODUCTION, INHERITANCE AND EVOLUTION

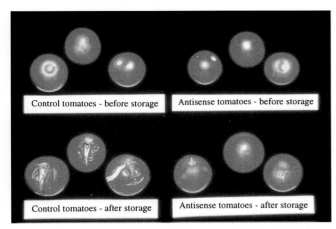

The tomatoes on the right have had their DNA altered so that they are more resistant to rotting. This technique (genetic engineering) is becoming more widely used (see Topic 4.17)

The human genome project

A technique called gene mapping can be used to identify the sites of genes on the chromosome. By the start of the twenty-first century, research now going on internationally may have finished working out the position of every gene on each of the human chromosomes, collectively the human **genome**. This knowledge should allow scientists to work out more accurately which children are at risk of inheriting particular diseases and could also enable cures to be found. It may also be possible to replace those genes that cause disease with 'normal' genes. Some companies are now patenting the genes they discover, hoping to recover the huge research costs of a project like this by charging others for using the benefits of their discovery. However, these moves have been controversial, both amongst research scientists and in the wider public. Research on human genes is subject to strict, ethical guidelines and legislation, to prevent exploitation.

★ THINGS TO DO

1 Observe cells that are dividing by meiosis. Draw and label what you see, giving the correct order to the stages.

2 Explain the difference between:
   a) a chromosome and a chromatid,
   b) diploid and haploid chromosome numbers,
   c) mitosis and meiosis.

3 **a)** Draw a large diagram to show how a cell with pairs of chromosomes will produce sex cells by meiosis. Include crossing over between chromatids in two of the chromosome pairs.
   b) Make another diagram to show how a condition like Down's syndrome might occur.
   c) The table shows the risk of having a Down's syndrome baby at different ages (data from *Higher Risk Pregnancy. Management Options*, ed. D K James, et al. (1995) W B Saunders & Co. Ltd, London, with kind permission).
   i) Draw a graph to show how the risk of having a Down's syndrome baby varies with age.
   ii) What does the graph tell you?
   iii) What advice about Down's syndrome might be helpful for newly pregnant women?

Maternal age/years	Risk of Down's syndrome baby
15	1 : 1578
20	1 : 1528
25	1 : 1351
30	1 : 909
35	1 : 384
40	1 : 112
45	1 : 28
50	1 : 6

4 **a)** Look up reproduction in plants and in animals and work out where (i) mitosis and (ii) meiosis will occur in each.
   b) Why doesn't meiosis occur in other cells?

5 Find out more about the human genome project, and write a report explaining the potential benefits and drawbacks of this work.

4.7 Plant enterprise

Jan and her husband Tom have a small plot of land. They grow fruit, vegetables and flowering plants for sale. To help keep their costs down, rather than buying new stock each year they provide their produce from their previous year's stock of plants. These are some of the methods they use to do this.

Flowering plants

Taking cuttings from good, healthy plants is a very cheap and effective way to produce plants. The new plants will be identical to the 'parent' plant that was cut, because the genes are identical. In the previous year Tom had taken stem and leaf cuttings from plants such as fuschias, geraniums, saintpaulias (African violets) and begonias. He had chosen the parent plants carefully to make sure they were strong growing and disease resistant. He had also noted which colours people preferred to buy, and chose most of his cuttings from those plants. The fuschia stem cuttings had been dipped in rooting compound and placed into compost. Kept in warm, moist conditions they rapidly developed new roots and began to grow strongly. Leaves of saintpaulia and begonia had been cut along the back and laid on trays of moist compost. Roots developed where the leaves had been cut and rapidly grew down into the compost. Tom also used some other methods to obtain new plants. The illustrations show different methods used with particular plants.

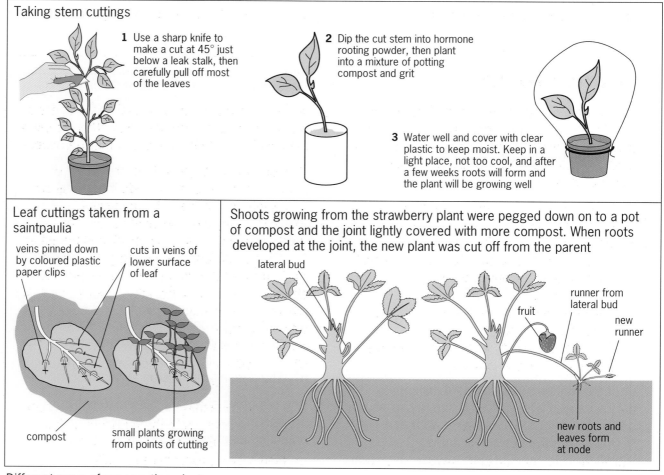

Different ways of propagating plants

REPRODUCTION, INHERITANCE AND EVOLUTION

In the vegetable plot

Jan had kept some of the potatoes (**tubers**) grown the previous year and stored them through winter (see Topic 2.2). As the weather began to improve, each one developed an eye – a point where a new stem would form. She cut the tubers up so that each piece contained an eye. In this way she managed to obtain several plants from each potato she had stored.

A potato sprouting new stems from the eyes

In the fruit beds

Many people had been keen to buy and grow their own strawberries. In the previous year, Tom had grown two or three different varieties. Some had cropped well. Others had been prone to diseases that caused the fruit to rot before picking.

He had kept those that had grown strongly, resisted diseases and produced good crops, and discarded the weaker, disease-prone ones. Towards the end of the season, each one had produced side shoots called **runners** from the base of the main stem. Small leaves and roots appeared from each runner at points called **nodes**. He placed the new plantlets in pots of compost and placed a small stone on top of the runner to hold it down. Within 2 weeks new roots began to form.

Eventually he cut the new plants from the runner away from the main plant and they grew on well. (See diagram opposite.)

★ THINGS TO DO

1 a) Taking cuttings from potted plants is quite simple. The diagram opposite shows how. If possible, investigate growing plants using this method. You might think of different things to investigate, such as:

- Is taking cuttings a reliable method of producing plants?
- What proportion of the cuttings actually succeed in developing into mature plants?
- Which types of plants can be reproduced by taking cuttings?
- What conditions are needed for the cuttings to root successfully?
- Are all of the plants grown from cuttings identical? How can you tell?

b) Imagine you have a nursery. What would you need to know about the method before starting to reproduce your stock? Make a list of questions and plan how they could be tested. Carry out your investigations (which may be shared with others in the class) and prepare a report, using both your own and other people's data, which could be useful to other gardeners.

2 New potato plants grow from the 'eyes' of potato tubers. These are buds that will grow into plants, given the right conditions.
a) Explain why the new plants should be identical to the potato that produced the tuber.
b) What conditions encourage good growth?
c) Cutting a tuber into small pieces will not stop the 'eyes' sprouting and growing into new plants. Think of ways to investigate how effectively they regrow. You might try to see how many plants you can grow from one tuber.
d) If possible carry out your investigation, and write a report of your findings.

3 Describe two features of asexual reproduction that are advantages to the gardeners. Now think of plants that can cause gardeners problems by reproducing asexually.

4.8 All the same

John was struggling to remove weeds from his vegetable garden.

Many plants, such as couch grass and bindweed, have developed ways of growing from small pieces. Some traditional methods gardeners use to produce new, identical plants were seen in Topic 4.7. Recently developed techniques can use much smaller pieces to grow new organisms.

NO MATTER HOW MUCH I TRY TO HOE OUT THESE COUCH GRASS WEEDS THEY STILL KEEP COMING UP.

THAT'S BECAUSE YOU'RE HELPING THEM TO SPREAD. THE HOE CUTS THROUGH THEIR UNDERGROUND STEMS, CHOPPING THEM INTO MANY SMALL PIECES. EACH PIECE CAN THEN GROW INTO AN IDENTICAL PLANT. SO HOEING IS ACTUALLY MAKING THINGS WORSE!

Cloning

Scientists can now copy nature on a much bigger scale. The technique they use is called cloning. It is a relatively simple, cheap and effective way of producing large numbers of identical plants. It has also been used to produce plants such as rare orchids, which are difficult to produce in other ways.

Cloning carrots

Carrots are one example of a plant that can easily be cloned. The process starts with a healthy carrot plant that is both disease free and has several qualities that a grower may want (taste, rapid growth, etc.). Each stage of the process shown in the diagram is carried out, taking care to reduce the risk of microbes growing and infecting the new plants.

Growing carrots by cloning

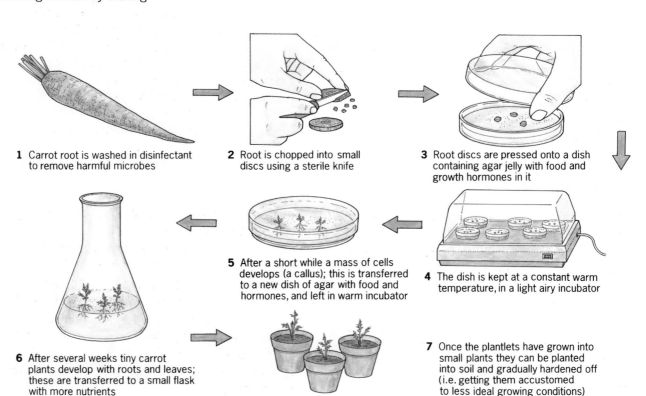

1 Carrot root is washed in disinfectant to remove harmful microbes

2 Root is chopped into small discs using a sterile knife

3 Root discs are pressed onto a dish containing agar jelly with food and growth hormones in it

4 The dish is kept at a constant warm temperature, in a light airy incubator

5 After a short while a mass of cells develops (a callus); this is transferred to a new dish of agar with food and hormones, and left in warm incubator

6 After several weeks tiny carrot plants develop with roots and leaves; these are transferred to a small flask with more nutrients

7 Once the plantlets have grown into small plants they can be planted into soil and gradually hardened off (i.e. getting them accustomed to less ideal growing conditions)

REPRODUCTION, INHERITANCE AND EVOLUTION

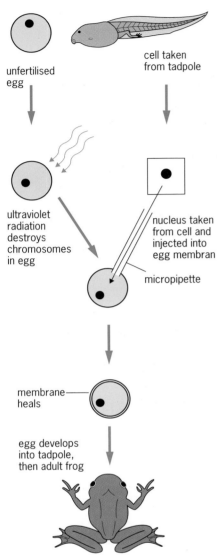

Cloning of frog cells to produce identical animals

Cloning animals

Scientists have been able to make identical copies of frogs and toads for some time. It is more difficult to clone animals than it is plants, but a wide range of animals can now be cloned, including sheep. The T cells (part of the immune system) of an embryo are used; they are separated at an early stage of development and then are transplanted into a 'host' mother. Here the cells develop into identical embryos, or clones.

Tissue culture

Animal cells, like plant cells, can be grown in a suitable liquid (a 'growth medium'), containing food and hormones, in dishes kept warm and aerated in incubators. The technique is called **tissue culture**, or **cell culture**. It starts with small pieces of organs from a freshly killed animal. Enzymes or corrosive chemicals are used to separate the cells. The cells receive all they need to stay alive from the growth medium. The solution is changed regularly to replace the materials used up by the cells and to remove harmful wastes. Great care is taken to avoid contamination of the medium with microbes. Chemicals made by the cells can be collected and purified.

A dish of tissue-cultured cells

Cells from mice and sheep can also be joined together (**hybridised**), then encouraged to produce multiple clone copies. These cloned cells are valuable because they make useful antibodies, which have a number of applications, including use in pregnancy-testing kits, purification of enzymes and diagnosis of certain infectious diseases.

★ THINGS TO DO

1 a) What are the advantages and disadvantages of cloning plants?
b) Cloning potato plants is more effective than growing them from tubers, but farmers generally still grow potato plants from tubers. Why do you think this is?

2 Tissue culture requires 'sterile techniques' to be used, to avoid microbes like bacteria contaminating the growth media.
a) What effect could microbes have?
b) What are the possible benefits of culturing lots of identical cells?
c) Freshly dead embryos from animals, and sometimes humans, have been used to start cultures of clones of cells. This is strictly controlled by government regulations. Consider the possible objections of using cells from embryos in this way. Try to find out more about animal tissue culture or cloning and write an essay discussing the rights and wrongs of using this technique.

4.9 More than a flower

There are millions of flowering plants on Earth. The flowers of some (e.g. grasses) are not always apparent. Others (e.g. some types of orchid) are so rare that people are willing to pay thousands of pounds for a single plant

The century plant (*Agave americana*) flowers only once every 100 years. In 1995, encouraged by the unusually warm weather experienced in Britain that year, one of these plants flowered at Kew Gardens. The flower lasted for only one day

Plant variation

The vast variety of flowering plants that exist today have evolved over millions of years. In each year, new plants have grown from seeds formed when pollen from one plant fertilises an egg cell (either in the same plant, or in another plant of the same species). The pollen carries genetic information from one plant. The egg cell contains genetic information from the other (unless the plant variety is self-fertilising). The resulting seed will have a mixture of characteristics from each of the 'parents' – it will produce a flowering plant that may be quite different to either parent.

Sexual reproduction in plants

Seeds begin to form when nuclei from **pollen** grains (the male sex cells in plants, equivalent to the sperm in animals) join with the egg cell (the female sex cell). The pollen grains are made in the male part of the flower known as the **anther** – part of the **stamen** (shown in the diagram). When the anthers are ready to release their pollen they split open and the pollen grains escape. They are then carried to another flower, either by the wind, or by visiting insects (as in the

Different offspring from sexual reproduction

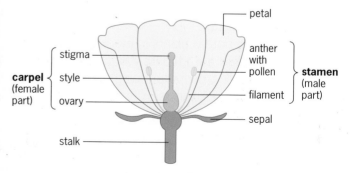

Section through a flower (buttercup)

REPRODUCTION, INHERITANCE AND EVOLUTION

A butterfly pollinating a flower

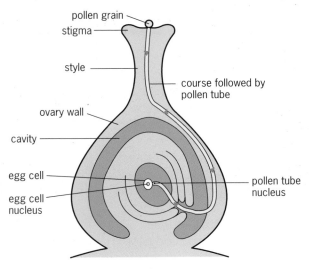

Growth of the pollen tube

photograph), where they stick to its **stigma**. The process of **pollination** is now complete.

When the pollen lands on the stigma it sends out a tentacle-like growth called a **pollen tube**. This grows into the stigma and down the **style**. Towards the tip of the pollen tube is a nucleus, which is equivalent to the nucleus in the head of an animal sperm. When the pollen tube reaches the ovary, it pushes its way through a small hole in the wall and grows towards the egg cell at the centre. When it meets the egg, the pollen nucleus fuses with the egg cell nucleus; the egg is then fertilised.

The fertilised egg divides repeatedly, and develops into a ball of cells, which is called the embryo. When fully developed, it is surrounded by special tissue called endosperm (or cotyledon(s)) that will supply food to it until it has germinated and is able to produce its food by photosynthesis (see Topic 2.4). The embryo and the tissue around it form the seed. The outer layer hardens into the protective seed coat (see the diagram in Topic 2.2).

★ THINGS TO DO

1 a) Collect ten different types of flower. Make a list of similarities and differences between each one. Why do you think they are so different?
b) Most plants have flowers with male and female sex organs. But some flowers only have one type. Suggest possible reasons for this.

2 Some plants have flowers that encourage cross-pollination (pollination by the pollen cells of other plants). Those of others favour self-pollination (i.e. by their own pollen cells).
   Is the plant in the photo adapted to favour self- or cross-pollination?

3 Your teacher will show you how to collect pollen grains and prepare a microscope slide so that you can look at them magnified.

a) Compare pollen grains from different flowers. Make a note of any features that might suggest whether they are adapted for cross-pollination or self-pollination.
b) Leave the pollen in a sample of sugary water to encourage it to develop a pollen tube. Observe and draw any tubes that grow.

173

4.10 Breeding to order

Producing new plants and animals, with the features that buyers want, is big business. Success depends on knowing what the best features are, and being able to reproduce enough of the species to satisfy demand at a reasonable cost. Topics 4.7 and 4.8 investigated different ways of reproducing plants asexually. But these methods do not produce new varieties; to get this, selective breeding methods are used.

MARIGOLD HHA **African Marigolds**
Superbly colourful, easily grown plants that are a must for a sunny border. Summer-long flowering performance will be achieved from a February–March sowing in gentle heat.
F1 Hybrid varieties
SUMO MIXED
12 09 80 An impressive blend of yellow, gold and orange flowers up to 9 cm (3.5 in) in diameter. The plants become smothered in bloom all summer through. Excellent garden performance and very uniform growth.
Ht. approx 30–40 cm (12–16 in). APC 50 . . . . 24H £1.99

Selective breeding in plants

F_1 varieties

Specialist seed nurseries often advertise 'F_1' varieties of both vegetable and ornamental flowering plants. Some are shown in the photograph. F_1 (for 'first filial (i.e. son) generation') varieties are selected that combine the best features of their parents. They are bred from parents chosen for their good features that 'breed true' for these features (i.e. they consistently produce offspring with the same features as themselves). Good features for vegetables usually include producing large numbers of crops, but may also include disease or pest resistance, or an extended season of cropping.

The F_1 carrot plant was developed by reproducing (i.e. crossing) parent plants as shown in the diagram on the right.

F_1 plants are hybrids, that is, they carry a mixture of the good features of parent plants (as genes). The parent plants carry two alleles for the chosen characteristic (see Topic 4.15), so one allele of each characteristic passes to the offspring.

Seeds of F_1 hybrid plants are more expensive than seeds produced by plants that have not been selected in this way, partly because they do not themselves breed true, so must be re-bred each year. But F_1 hybrids combine the good qualities of the parents and are usually high yielding. So the extra food produced pays for the higher cost of the seeds.

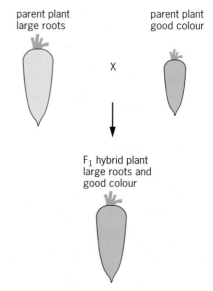

How F_1 varieties are bred

Modern food plants

More and more food plants are F_1 plants. Modern wheat plants are an example. Improved qualities that have been interbred include: better resistance to disease and cold conditions, shorter plants but with bigger ears containing more grain, the ability to germinate earlier and so grow for longer, and a higher protein content.

Selective breeding can also increase the size of plants because of the effect known as 'hybrid vigour'. If the offspring receive a combination of alleles that increase growth, the combination may not simply add the effects, but multiply them. For instance, maize (sweetcorn) plants can be bred to produce F_1 hybrids that have much bigger fruit (cobs) than their parents, thus greatly increasing the yield.

REPRODUCTION, INHERITANCE AND EVOLUTION

Selection in animals

Selective breeding is not restricted to plants. Modern farm animals are the result of selective breeding, using essentially the same technique as described above. Cattle are bred for their ability to produce milk or for good meat. Different varieties of hen are bred for egg laying or as chickens for eating.

Selection of an animal depends on many factors. For beef cattle this can include:

- the potential body size,
- the type of meat produced on the body (lean or fatty),
- how likely calves are to survive,
- how quickly calves grow into adults,
- how efficiently food is converted into meat.

By crossing parents with the preferred qualities, calves are born that suit farmers' needs.

Two types of beef cattle:
(top) Herefords; (bottom) Charolais

★ THINGS TO DO

1 Our understanding of inheritance dates back to the work of an Austrian monk Gregor Mendel, in the mid nineteenth century. He discovered that peas with different characteristics passed on these features to new pea plants systematically. Find out what Mendel discovered, and why his work took 40 years to affect the thinking of other scientists.

2 a) Seeds from F_1 hybrid plants (called 'F_2' plants) show a variable range of characteristics, unlike their F_1 parents. Why is this less desirable commercially?
b) How can the inheritance of alleles from the F_1 hybrid parent explain F_2 variability?

3 The table below shows recent changes in the number and status of different types (breeds) of cattle in western Europe.
a) Comment on the variation in breeds.
b) A smaller number of breeds reduces the number of alleles available for selective breeding. Why is this a problem for the breeder?
c) Why should some animals be kept of breeds that are unsuitable for modern farming needs?
d) Why is international co-operation needed to retain biodiversity (i.e. different breeds)?
e) Why may protecting species in the wild benefit future selective breeding programmes?

Region	Major	Minor	Rare	Status	
				Nearly extinct	Recently extinct
GB	10	4	8	3	2
France	7	3	6	5	2
All W. Europe	36	26	30	14	9

Total population size for all breeds in W. Europe in 1993 = 124 780 000
(Data from *Managing Global Genetic Resources. Livestock* (1993), Committee on Managing Global Genetic Resources Agricultural Imperatives. National Academic Press, Washington DC, with kind permission.)

4.11 Human reproduction

A new birth

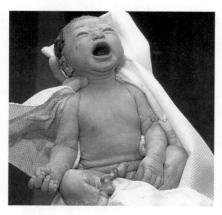

A happy day, some 9 months/39 weeks/270 days after conception

Pavendra explains how her first child was born:

The first sign was a bad pain in my back. Then my waters burst – the amniotic sac had broken. I went to hospital – quickly!

The pains got worse as muscles in the womb started to push on my baby. After 12 hours the nurse said that the opening of the womb (cervix) was nearly wide enough for the baby's head to push out. The pain was hard to bear, but soon over. The cervix widened to 10 cm and with some extra shoves the baby was forced out.

The shock of being cooler outside my body and having to breathe air made her cry out – a wonderful sign of life!

The birth of a baby is a wonderful thing, but it has been alive for many months already inside its mother's womb during her pregnancy. It was formed when a sperm from the father joined with an egg from the mother; this is fertilisation or **conception**. Humans, like other mammals, produce offspring (babies) by a type of sexual reproduction in which sperm and egg are brought together inside the female's body ('internal fertilisation'). This gives a better chance of them meeting, so conception is more likely to take place.

Taking advantage of nature's way

Once a female becomes sexually mature (at puberty), one egg is released from an ovary every month. A series of changes take place inside the uterus, ending with a discharge of blood (a menstrual period). These changes (the **menstrual cycle**) are repeated each month until conception. A period does not usually happen after an embryo has started to form – so a missed period is one sign of pregnancy. The cycle is controlled by female sex hormones, chemicals that tell parts of the body what to do (see Topic 4.13).

An egg is normally released around the 14th day of the menstrual cycle, and stays alive for up to 36 hours. Sperm can survive for 3 days. Love making just before or after release of the egg gives the best chance of egg and sperm surviving long enough to meet, and for conception to take place. So at this time of the month the female has a greater chance of becoming pregnant. Couples not wanting to conceive should be especially careful then.

When a couple make love they share physical and emotional experiences. They also share a responsibility to each other, and to the new baby that could result. Pregnancy can be a wonderful condition, for both mother and father.

During love making both partners become sexually aroused. The male's penis stiffens as blood enters it (erection). Continued movement of the penis inside the vagina leads to sperm being squirted into the top of the vagina – this is ejaculation.

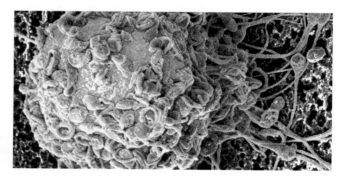

An egg, surrounded by sperm at the moment of conception: the egg is being fertilised by just one of these sperm

REPRODUCTION, INHERITANCE AND EVOLUTION

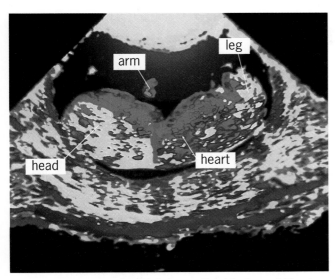

Ultrascanning (see *GCSE Science Double Award Physics*, Topic 4.4) is a technique that is used to see the baby as it develops inside the womb. It causes no pain or suffering to mother or child, and can show how well the baby is developing. It can also show the baby's sex

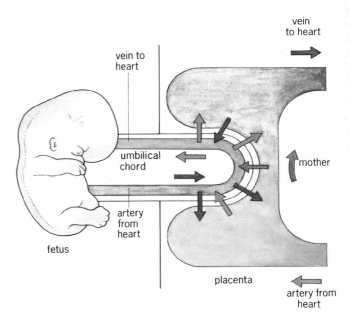

Exchange of substances across the placenta

In the womb

The fertilised egg develops inside the mother's womb, where it is warm, safe and well provided with all its needs to survive and grow well. Once an embryo has implanted on to the uterus wall the point of attachment develops into the **placenta**. This is an area where the blood capillaries from the mother almost meet those from the embryo. Materials and heat can then be exchanged between them, as the diagram shows.

The embryo quickly develops a heart to pump blood around its body. It forms human features and becomes known as a **fetus**. The fetus is protected from physical damage and kept warm by the surrounding liquid, the **amniotic fluid**. The placenta prevents most harmful chemicals and germs entering the fetal blood from the mother. Altogether, the fetus has almost ideal conditions in which to grow. So when it is time for birth, the baby will be in for a shock!

★ THINGS TO DO

1 Look at the photograph of an egg and sperm.
  a) What is the job of each cell?
  b) Describe the structure of each.
  c) Explain how the structure of each allows it to do its job well.

2 a) Newly pregnant females are advised not to smoke or drink much alcohol, to avoid harming their developing child. How could alcohol and nicotine from cigarettes reach the fetus?

  b) Why could these substances be more harmful to the fetus than to the mother?
  c) What other precautions should a mother to be take so that her baby develops safely?
  d) How should she modify her diet?
  e) Collect information on pregnancy and make a poster to show what mothers and fathers to be can do to prepare for the birth of their child.

4.12 Planned pregnancy

It's not always easy

Jan and Dave have been trying for several years to have a child, but without success. They've decided to ask a doctor for advice.

YOU'RE MAKING LOVE OFTEN ENOUGH AND AT THE RIGHT TIME, AND NOT USING CONTRACEPTIVES, SO YOU HAVE A GOOD CHANCE OF CONCEIVING. BUT REMEMBER, IT'S STILL ONLY A CHANCE. VERY OFTEN, COUPLES WANT CHILDREN SO MUCH THAT THEY DON'T GIVE NATURE LONG ENOUGH TO ACT. IT'S QUITE A FEAT FOR AN EGG TO BE FERTILISED. I'LL EXPLAIN WHAT I MEAN.

HERE ARE THE MALE SEXUAL ORGANS. EACH TESTIS MAKES MILLIONS OF SPERM EACH DAY. THE SPERM ARE STORED INSIDE THE SPERM TUBE. THEY ARE SQUIRTED OUT OF THE PENIS WHEN THE MAN AND WOMAN MAKE LOVE — THIS IS EJACULATION. BUT SOMETIMES FEWER SPERM THAN NORMAL ARE MADE, AND ALMOST ALL DIE WITHOUT GETTING NEAR TO AN EGG. SPERM CAN LIVE ONLY ABOUT THREE DAYS.

Male — sperm tube, penis, testis

Female — thick uterus lining, oviduct, path of egg, sperm, egg, uterus, cervix, vagina

THE SPERM SWIM TO THE WOMB (UTERUS) INSIDE THE WOMAN'S BODY, THEN UP INTO THE OVIDUCTS. EACH MONTH AN EGG IS RELEASED FROM ONE OF THE TWO OVARIES. THE EGG MOVES DOWN THE OVIDUCT TOWARDS THE UTERUS. IF ANY SPERM FIND THE EGG, THE HEAD OF ONE SPERM ENTERS AND FERTILISES IT. FERTILISATION HAS TAKEN PLACE. BUT SOMETIMES AN EGG IS NOT RELEASED, OR IT DIES BEFORE ANY SPERM CAN REACH IT, SO THERE IS NO CONCEPTION.

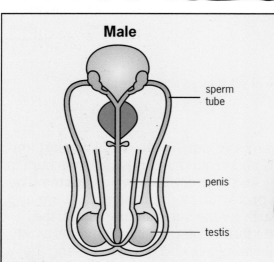

fertilised egg dividing as it is moved along
egg descending + sperm around
oviduct
thick lining (wall) of uterus
cervix
'ball of tiny cells' embryo attaching to wall of uterus (womb)

THE FERTILISED EGG DIVIDES INTO A BALL OF THOUSANDS OF CELLS — AN EMBRYO. TO SURVIVE, IT HAS TO STICK TO THE WALL OF THE UTERUS — TO BECOME EMPLANTED. THEN IT CAN RECEIVE THE FOOD, OXYGEN AND OTHER CHEMICALS IT NEEDS FROM ITS MOTHER'S BLOOD. A PLACENTA DEVELOPS SO THAT THESE CHEMICALS CAN PASS EASILY TO THE EMBRYO AND TAKE AWAY HARMFUL WASTES. FERTILISED EGGS DON'T ALWAYS IMPLANT, SO SOME DIE.

Fertility drugs

Jan has been told that her ovaries are not releasing eggs normally. In fact, it's likely that eggs are not being made. A possible treatment is with fertility drugs containing female sex hormones. A particular hormone called **follicle-stimulating hormone** (**FSH**) (see Topic 4.13) is used to treat infertility because it stimulates the development of **follicles** (groups of cells in the ovary which make an egg). So there's a much better chance that eggs will be released and fertilised. Each ovary has thousands of cells that could develop into follicles and grow eggs inside. But their development depends on FSH acting as a trigger. If the natural level of FSH inside a female's ovaries never reaches the trigger point then follicles do not develop.

You must be sure that you want to use a fertility drug as there can be drawbacks. It is difficult to calculate the precise dose of FSH needed, and sometimes fertility treatment can cause many eggs to be released at once, leading to multiple births.

Another method used is **in-vitro fertilisation**, which involves taking the eggs out of the ovary to fertilise them and then putting them back, but it's a more difficult and costly treatment.

In-vitro fertilisation technique

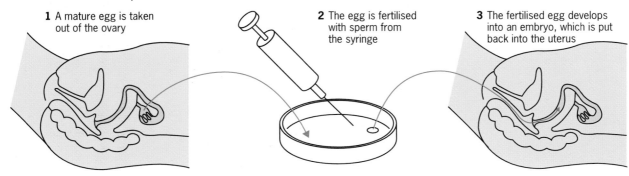

1 A mature egg is taken out of the ovary
2 The egg is fertilised with sperm from the syringe
3 The fertilised egg develops into an embryo, which is put back into the uterus

★ THINGS TO DO

1 Look at the headlines below. Think of the benefits and drawbacks of using hormones to control fertility. If possible, discuss your views with others and find information from other sources.

Write an article for a magazine to explain your ideas.

> A mother at last! After 10 years trying for children, Jackie gives birth thanks to fertility drug treatment

> Six of the best! Susan Smith's fertility treatment ends in sextuplets

> Concern about side-effects from fertility drugs

2 a) Make a list for Jan and Dave of things that might prevent an egg being fertilised.
b) Explain why making love 1 day after an egg is released could result in conception.
c) Why is conception unlikely when love making occurs 7 days before an egg is released?
d) Look at the graph. Why should Jan and Dave be more concerned about Jan not conceiving if they were older?

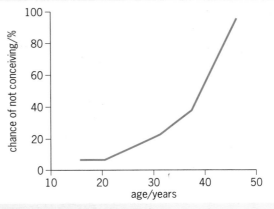

Chances of not conceiving with age

4.13 Controlling fertility

Mandy is learning about the different ways to prevent pregnancy (**contraception**). Her notes are shown on the right. Effectiveness rates are shown for certain methods of contraception. Some are more effective than others. The method chosen by a male or female will depend on, for instance, what they prefer, their religious beliefs, their social background and whether they will want children in the future. Whatever decision they make, it should be taken before love making: just hoping that pregnancy will not happen is *not* a form of contraception.

Contraceptive hormones

Mandy has decided to use a contraceptive pill and is asking the doctor for advice.

I THINK YOU SHOULD USE THIS TYPE OF CONTRACEPTIVE PILL TO PREVENT PREGNANCY. IT IS A MIXTURE OF FEMALE SEX HORMONES THAT STOP EGGS BEING MADE. IT IS A VERY EFFECTIVE FORM OF CONTRACEPTION, BUT A TINY MINORITY OF WOMEN SUFFER SIDE-EFFECTS. COME BACK IF YOU HAVE ANY PROBLEMS.

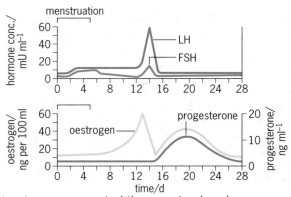

How hormones control the menstrual cycle

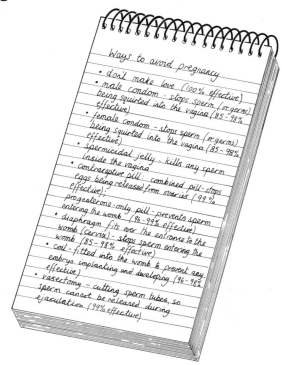

Forms of contraception

Female sex hormones

Events in the menstrual cycle are controlled by sex hormones. Four hormones are involved, as shown in the diagram opposite:

- **follicle-stimulating hormone** (FSH) and **luteinising hormone** (LH), both made in the pituitary gland underneath the brain,
- **oestrogen** and **progesterone**, both made in the ovaries.

Oestrogen is the name given to a group of very similar female sex hormones that have similar jobs. Contraceptive pills contain oestrogen, and sometimes progesterone. Oestrogen is used to suppress the release of FSH from the pituitary gland, so follicles do not develop inside the ovaries. However, high levels of oestrogen in the blood can have side-effects, such as increased weight and a slightly greater risk of developing certain cancers or thrombosis. These risks must be balanced against the risks of pregnancy itself, and the benefits to lifestyle and women's health from having smaller families.

REPRODUCTION, INHERITANCE AND EVOLUTION

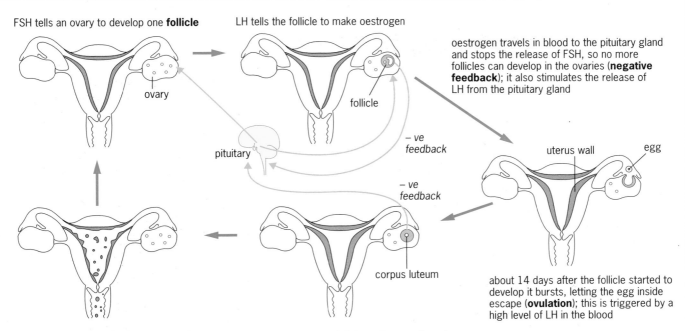

How hormones control the menstrual cycle

★ THINGS TO DO

1 a) Make a table to show each type of contraception, with its advantages and disadvantages.
b) Why should people seek advice before using contraception?

2 The graphs opposite show how hormone levels vary through a menstrual cycle when conception does not take place. Copy them and indicate where:
a) FSH stimulates follicles to develop,
b) a rise in FSH causes a burst follicle to change into a corpus luteum,
c) a sudden drop in oestrogen stimulates LH and FSH release from the pituitary gland,
d) LH triggers ovulation,
e) progesterone causes thickening of the uterus wall,
f) a drop in progesterone level allows a period to start.

> Male workers in the contraceptive pill factory are growing breasts
>
> Doctors advise women to change their brand of pill because of thrombosis risk
>
> Is China's one-child policy failing? Fears of grain shortages
>
> Women in developing countries say the pill gives them a better lifestyle
>
> Church leaders vote against the pill

3 Look at the headlines above. They are typical of the kinds of issues that the contraceptive pill raises in societies around the world. Look for more information on this.
a) Make a table, listing the various points for and against its use. Include the views of the women (and men) involved, and the wider society. Weigh up the sizes of the risks and benefits.
b) Who do you think is right? Explain why.

181

4.14 Passing it on

Twins

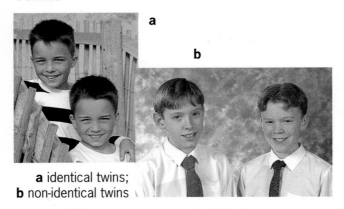

a identical twins;
b non-identical twins

Identical twins have the same set of chromosomes in the nucleus of each of their cells. They have inherited exactly the same information from their parents because, just after fertilisation, the fertilised egg split into two identical cells, and each then developed into an embryo. Non-identical twins result from two separate eggs fertilised by separate sperm. Because these two fertilisations happened at about the same time, the two embryos developed and were born together. But each carries different sets of chromosomes, so the twins are not identical, and in fact no more alike genetically than other brothers and sisters born at different times.

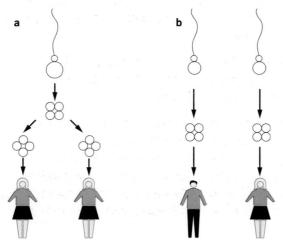

How fertilisation takes place in: **a** identical twins; **b** non-identical twins

Boy or girl?

The sex of a baby is determined by two particular chromosomes out of the 46 in the cell; these are the sex chromosomes. One sex chromosome is carried in the nucleus of the sperm (so is inherited from the father). The other is inside the egg nucleus (so is inherited from the mother).

The baby is female if an X chromosome joins up with another X chromosome at fertilisation. A male results when an X chromosome pairs up with a Y chromosome. The diagram shows how this happens. (For details of meiosis see Topic 4.6.)

The chance of a baby being a girl or boy is two out of four. This is an even chance (or a 50 : 50 chance).

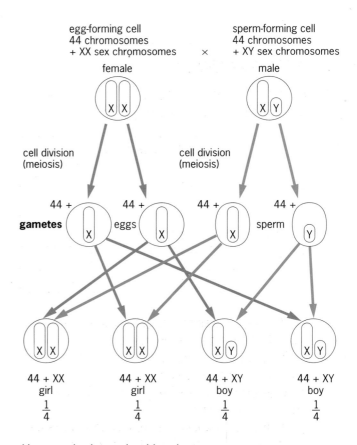

How sex is determined by chromosomes

A male condition

Some men cannot do jobs that women can. It's not a question of sexual discrimination, but the result of inheritance. An example of this is colour blindness; it is a condition that affects about 8% of males, but is very rare in females. Most males who suffer from it cannot distinguish between red and green colours, since both appear as the same shade of grey.

Colour blindness is inherited. The cause is a gene carried on the X chromosome. But this usually has the effect of causing colour blindness only when it comes together with a Y chromosome. So it is males that suffer, as the diagram shows. A female with the colour-blind information in her genes is called a 'carrier' because although not colour blind herself, she will pass the information on to any children she has. Her sons will then be colour blind.

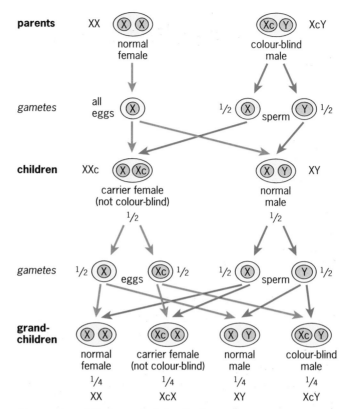

How colour blindness is inherited

★ THINGS TO DO

1 The data show the number of girls and boys born in one city over several years.

Year	Number of live births	
	Male	Female
1990	667	702
1991	598	612
1992	633	571
1993	772	748
1994	546	578

a) Calculate the percentages of boys and girls respectively born each year.
b) Calculate the average percentages of boys and girls born over 5 years.
c) What proportions of boys and girls would you expect to be born?
d) What might explain any difference between your expectation and the actual figures?
e) Doctors can now tell the sex of a child before birth. Scientists think it would be possible to arrange for a male or female baby to be conceived artificially. Think about the benefits and drawbacks of parents deliberately deciding the sex of their children. Discuss your ideas with others, and write a short article for a magazine.

2 Explain why identical twins must be the same sex, but non-identical twins can be either the same or different sexes.

3 Some jobs, like driving trains, cannot be done by people with colour blindness.
a) What other jobs might be difficult or dangerous for someone with colour blindness?
b) A colour-blind female has sex chromosomes X_cX_c. Explain how these could be inherited.

4.15 Inheriting features

Different features, or 'characters', are passed from parents to offspring as *genetic* information – that is, genes on chromosomes. Some features are controlled by single genes; some result from a combination of genes. Still others are the effect of the *environment*; in humans and other higher animals, these will not be inherited. An example of an inherited character determined by a single gene is tongue-rolling ability.

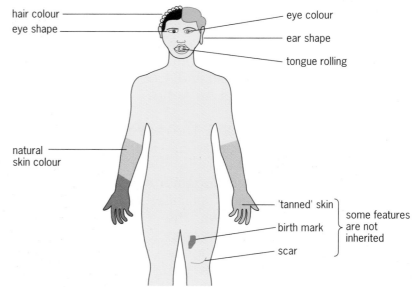

Inherited and non-inherited features

Tongue rolling

The ability to roll the tongue depends on one pair of alleles (different forms of a gene). Having just one copy of the allele for 'tongue rolling' gives the person the ability to roll their tongue. This allele is therefore said to be **dominant** because it seems to be stronger in effect, and to 'overrule' the other form of the gene ('no tongue-rolling ability'). The dominant form is usually represented by a capital letter (e.g. R).

The other form of this gene lacks the information needed to make the tongue roll. This allele is not dominant, as two copies of it are needed before the characteristic is seen. It is said to be **recessive**. It is represented by a lower-case letter (in this case by 'r') to show its effect is less strong.

The diagram shows the different possibilities for inheriting the ability to roll the tongue.

The diagram showing possible combinations of alleles explains the *pattern* of inheritance of characteristics like tongue rolling. It also shows the *possible chance* of a baby showing a characteristic of the parent.

When a pair of chromosomes in a cell both carry the same alleles, the individual is said to be **homozygous** for that characteristic. So a tongue roller containing the alleles RR is homozygous. A non-roller, with the combination rr, is also homozygous.

When a pair of chromosomes in a cell carry two different alleles, the individual is said to be **heterozygous**. So a tongue roller carrying the alleles Rr is heterozygous.

A tongue roller may have either the RR or the Rr combination of alleles. To find out which, you would have to look at their children.

Some people can roll their tongue, but others cannot; this ability is inherited

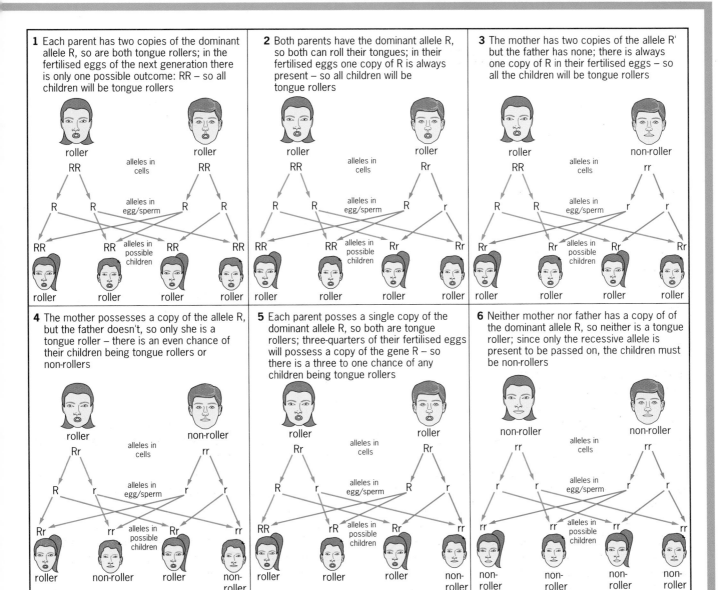

Inheritance of tongue-rolling ability

★ THINGS TO DO

1 In some countries of the world people have large families. Try to explain what alleles for tongue rolling the parents have in the following families (T = tongue roller; NT = non-tongue roller):
 a) family A; children are T, T, NT, T, NT, NT, NT, NT,
 b) family B; children are NT, NT, NT, NT, NT, NT,
 c) family C; children are T, T, T, T, T, T, T, T, T, T.

2 a) What is meant by being homozygous?
 b) What is meant by being heterozygous?
 c) Two parents, the mum a non-tongue roller, have one child who is a tongue roller. Explain why the dad could be either homozygous or heterozygous for tongue rolling.

4.16 What's the chance?

It is thought that there may be as many as 100 000 genes in the nucleus of a human cell. Some of these are responsible for the 3000 or so diseases, like cystic fibrosis, that are known to be inherited.

Cystic fibrosis is the commonest life-threatening inherited disease in Britain. About 6000 people suffer from it, and about 1 in 20 people are healthy carriers of the recessive allele that causes it. If a person's cells possess two copies of the recessive alleles it means that their glands release excessive amounts of mucus. This sticky substance collects in the lungs, making breathing difficult and increasing the risk of infection. It also prevents effective digestion of food, resulting in malnutrition. Currently there is no cure, but the disease symptoms can be made less severe by physiotherapy, exercise, drugs and appropriate diets. Now that the gene causing it has been identified, people at risk of it are offered 'genetic counselling'.

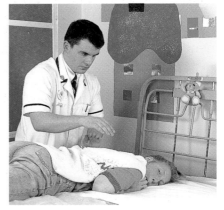

A cystic fibrosis sufferer having physiotherapy

A genetic counselling session for cystic fibrosis

WE NEED ADVICE BECAUSE LOUISE IS PREGNANT AND KNOWS THAT SHE IS A CARRIER FOR CYSTIC FIBROSIS. WE'RE WORRIED THAT OUR CHILD MIGHT SUFFER FROM IT.

YOU'VE DONE THE RIGHT THING TO SEEK ADVICE. CYSTIC FIBROSIS IS CAUSED BY A RECESSIVE ALLELE. SO BOTH RECESSIVE ALLELES MUST BE PRESENT FOR THE DISEASE TO SHOW. LET'S CALL THE NORMAL (NO DISEASE) ALLELE C AND THE CYSTIC FIBROSIS ALLELE c. THE POSSIBLE COMBINATIONS OF ALLELES ARE: (i) CC (NORMAL); (ii) Cc (NORMAL BUT A 'CARRIER') AND (iii) cc (CYSTIC FIBROSIS). YOU KNOW THAT ONE OF YOU (LOUISE) IS A CARRIER (THAT IS, SHE HAS A COPY OF THE ALLELE THAT CAUSES THE DISEASE). BUT YOU DON'T KNOW ABOUT THE OTHER (JIM). LET'S SEE WHAT THE CHANCES ARE OF YOUR CHILD INHERITING THE DISEASE. THERE ARE TWO POSSIBILITIES:

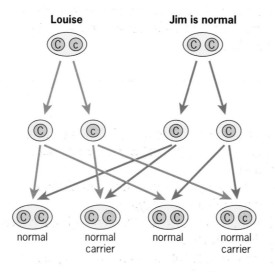

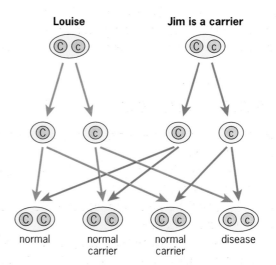

REPRODUCTION, INHERITANCE AND EVOLUTION

Other inherited diseases

Huntington's chorea

Huntington's chorea is a very rare disease of the nervous system. It is caused by inheritance of a dominant allele, so only one copy of the allele is needed in an individual for the disease to show.

If two people who are heterozygous carriers for the disease have children, there is only a one in four chance that any child will be normal, as the diagram shows.

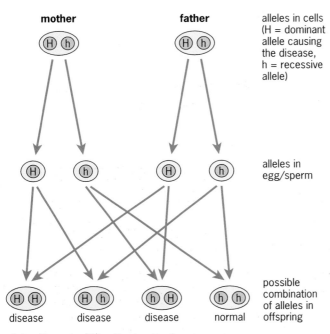

Inheritance of Huntington's chorea

Symptoms of the disease do not become obvious until middle age, by which time a sufferer may already have passed on the gene to their children. However, a technique called 'genetic marking' can be used to identify whether the allele is present in people thought to have possibly inherited the disease.

Sickle cell anaemia

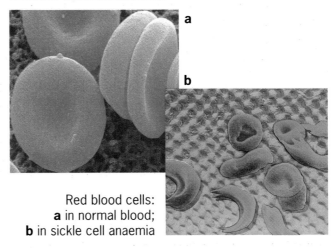

Red blood cells:
a in normal blood;
b in sickle cell anaemia

Sickle cell anaemia is an inherited disease that changes the shape and effectiveness of red blood cells. The gene affected is responsible for making a substance called globin, an essential part of haemoglobin, which carries oxygen. Like cystic fibrosis, the disease is caused by a recessive allele, so it develops only in people who have two copies of the allele (i.e. are homozygous for it). However, carriers of the disease tend to suffer less from a blood disease called malaria, carried by mosquitoes. So it seems that in this case there is actually an advantage to being a carrier, particularly in countries where malaria is a common life-threatening disease.

★ THINGS TO DO

1 a) Explain what is meant by the term 'carrier of cystic fibrosis'.
b) Explain why it is not possible to be a healthy carrier of Huntington's chorea.
c) How can it be an advantage to be a carrier of sickle cell anaemia?
d) Can you find any other examples of inherited diseases? Write down what is at present known about their causes.

2 Show what combinations of alleles can be produced when the following parents have children:
a) mother homozygous for Huntington's chorea, father heterozygous,
b) father homozygous for Huntington's chorea, mother heterozygous,
c) mother and father both homozygous for Huntington's chorea.

4.17 Changing the message

Genetic engineering in transplants

Transplanted organs, such as heart and kidney (see Topic 1.18), work well when the body accepts the new, donated organ. But transplantation sometimes fails; the new organ can be rejected because the person's immune system recognises the organ as different tissue to its own. This problem can be minimised when human transplant organs are used (particularly from members of the same family). Organs from animals are more readily available, but are usually rejected as 'foreign tissue'. However, recent research has produced great hopes that organs from pigs can be successfully transplanted into humans. These donor pigs are no ordinary pigs though, but have been **'genetically engineered'** so that their organs contain human DNA; this gives cells in the animal's organs a coating of human protein molecules. The person's body is therefore fooled into accepting the transplanted organ as being human.

Bacterial production of insulin

Many people suffering from diabetes (see Topic 1.22) lead normal lives by regularly injecting a hormone called insulin into their blood. Bacteria now make much of this insulin. It is of no use to the bacteria; they have been altered, by genetic engineering, so that they make a product purely for human consumption. Genetic engineering involves transferring the gene for making human insulin from a chromosome in a human cell into a bacterium. The diagram outlines how this is done.

Bacteria can be grown efficiently and cheaply in large fermenters. These are automatically controlled to ensure production of high-quality insulin, made 24 hours a day, 365 days a year. One alternative to bacterial-made insulin is to extract insulin from the pancreas of dead pigs. This is less effective, and the insulin is also of inferior quality.

1. Pancreas cell containing the gene (a length of DNA) for making insulin; DNA makes RNA, which then makes insulin

2. The RNA is used as a 'template' (plan) to remake the length of DNA that is the insulin gene; this process needs an enzyme, called reverse transciptase

3. The DNA 'insulin' is inserted into a circular loop of DNA (a 'plasmid') from a bacterium

4. This plasmid is then transferred into a bacterium

5. The bacterium divides, passing copies of the insulin gene into each new bacterium; further cell divisions produces clones of bacterial cells that are all able to make human insulin

Genetic engineering of insulin

A fermenter at work, producing human insulin

REPRODUCTION, INHERITANCE AND EVOLUTION

Uses of genetic engineering

Transferring genes from animals to humans, or humans to bacteria, can bring many benefits. Bacteria containing human genes can make a variety of chemicals and hormones, like insulin and growth hormone, for use as drugs to treat major human diseases.

Genes can also be transferred from plant to plant, from plant to bacterium, or from bacterium to plant. Gene transfer can improve plants in many ways (e.g. better resistance to disease or poor weather, quicker growth, improved flower colour, longer-lasting fruit). Bacteria with genes transferred from plants can make products that are difficult to make by other methods.

Gene transfer may be used in humans in the future to treat diseases (see 'Human genome project' in Topic 4.6).

★ THINGS TO DO

1 Read the newspaper article, and try to find more information about the current and future uses of genetic techniques. Discuss your ideas with others, then write a short essay, comparing the benefits and drawbacks of these methods.

DNA fingerprinting allows police to identify suspects from the DNA structure of their genes

2 Discuss the relative merits of these different techniques:
   a) insulin produced from genetically engineered bacteria, or extracted from dead pigs,
   b) human growth hormone made from bacteria, or removed from freshly dead people,
   c) blood transfusion using artificial blood with haemoglobin from genetically engineered bacteria, or from human donors,
   d) transplants of organs from genetically engineered pigs or those from human donors.

3 Scientists are trying to find new ways of preventing fruit like tomatoes from rotting too quickly. One technique being used (genetic engineering) is to insert genes that stop the production of ethylene by tomato cells.
   a) What are the benefits of this technique?
   b) What is genetic engineering?
   c) Find out what laws apply to producing food from genetically changed plants.

Genetic tests should be monitored

Leading scientists are calling for strict rules on how genetic testing is carried out. One professor said that, if carried out properly, genetic testing and engineering could benefit thousands of families afflicted by inherited disease. Postal tests are now available to identify carriers of common single-gene defects such as cystic fibrosis and muscular dystrophy. Screening for susceptibility to certain cancers, including bowel cancer and some breast cancers, is also based on people's genetic make-up. It is becoming increasingly possible to predict a person's risk of developing disease as a result of genetic and environmental interactions in cases such as diabetes, asthma and rheumatoid arthritis.

But dangers exist of potential abuse and ignorance. If do-it yourself genetic testing became widely available it could cause more harm than good by generating widespread anxiety over genetic predispositions about which nothing could be done. People identified as having a susceptibility to life-threatening conditions, such as heart disease, might find insurance difficult to buy. Records of an individual's genetic make-up could also be misinterpreted or deliberately used against them by unscrupulous people.

Many people think that it is morally wrong to patent genes. They fear that such legal moves will reduce the free flow of research information and hold back vital new discoveries.

4.18 Mutation

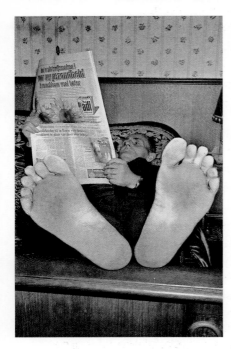

This person has six toes. These features are due to the person's genetic make-up; the genes responsible were inherited from the parents, but have changed (mutated) in the process, producing these abnormal features

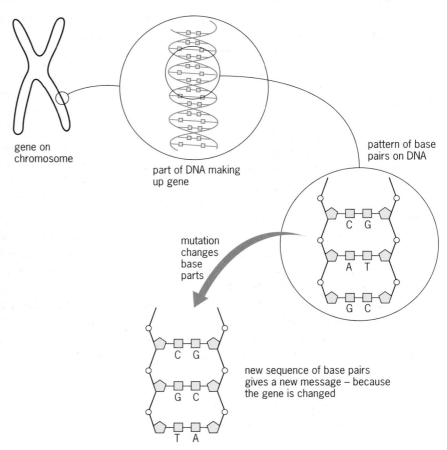

How a mutation happens

Natural and induced mutations

A **mutation** is a change in the DNA from which a gene is made. The new version of the DNA is then inherited, as a different allele. Mutations can occur naturally; there is on average a one in six million chance of a gene mutating. But the risk of this happening is increased by exposure to:

- ionising radiation from the sun (ultraviolet light), X-rays, or radioactive substances such as radon gas (given off by rocks such as granite) (see *GCSE Science Double Award Physics*, Topic 4.14),
- certain chemicals.

Chemicals and radiations that increase the mutation rate are called **mutagens**: the greater the dose of a mutagen, the bigger is the chance of mutation. This effect can be either useful or very harmful, or it may have no obvious effect.

Beneficial mutations

Microbes and fungi can be changed by mutation to make them more effective producers of useful materials such as **antibiotics** (drugs that kill bacteria). After exposure to the mutagen, those with the desired properties are selected, and these are then cultured. The process is easy, cheap and successful.

The antibiotic penicillin is made by the fungus *Penicillium chrysogenum* to kill its competitors, which are certain types of bacteria. It grows naturally on the surface of fruit (see Topic 3.12). Repeated exposure of the fungus to mutagens such as mustard gas and gamma radiation in controlled conditions produces new strains that make much more penicillin than the original type. By selecting these strains and growing them on, high yields of penicillin become possible and the cost of production drops.

REPRODUCTION, INHERITANCE AND EVOLUTION

Harmful mutations

Drug companies continually make new kinds of antibiotic. This is necessary because each antibiotic is effective against only a small range of bacteria. Bacterial reproduction and mutation produce a continual variety of slightly different types (strains), some of which will be naturally resistant to the antibiotic (in other words the drug does not affect them). These drug-resistant strains survive while others die. The surviving resistant strain then reproduces quickly and becomes widespread, making the antibiotic worthless, and the eventual risk to people much greater. This is an on-going example of selection (see Topic 4.3): the resistant strain has been 'selected' because it has features that allow it to survive, whilst others die out.

Leukaemia is a form of cancer that affects the white blood cells (leucocytes); they become abnormal, and grow and divide in an uncontrolled way. It results from a change in the structure of two chromosomes, with an exchange of chromosome bits. This is called **translocation**, and is shown in the diagram. The likelihood of this happening increases dramatically with exposure to mutagenic agents.

Most mutations are harmful. Workers exposed to mutagens may suffer if:

- reproductive cells are affected – mutations in these cells produce gene alterations, which could be passed on to their children, showing up as abnormal development, or even death, of the developing embryo in the womb,
- mutations occur in body cells, causing uncontrolled cell division and growth – i.e. cancer; these abnormal cells may then spread, invading other parts of the body.

Gene mapping (see Topic 4.6) can be used to locate the site of a gene causing disease, such as that responsible for muscular dystrophy. Cloning of this gene to produce multiple copies gives researchers the opportunity to study the DNA pattern of the normal gene and see how mutation has changed it. Finding a way of 'correcting' that gene, or removing its harmful effect, offers a possible cure.

a
no. 9
no. 22
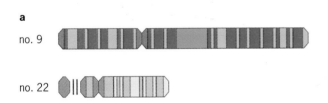

b
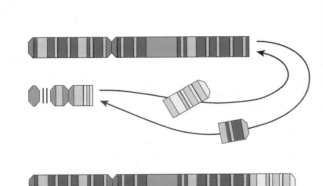

c

The chromosomal rearrangement (translocation) that causes leukaemia: **a** normal chromosomes 9 and 22; **b** translocation; **c** translocation chromosomes

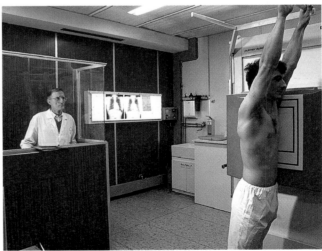

Some people, like this radiographer, work with mutagens (in this case X-rays and radioactive materials) every day. They minimise the risk to their health by taking precautions such as standing behind lead shielding, which blocks the harmful rays

MUTATION

Cancer and EMF

Parents who blame high-voltage power cables for leukaemia and brain tumours in their children received a boost to their claim for compensation today with the first explanation of how electromagnetic fields may trigger cancer.

Scientists at Bristol University have evidence that overhead power lines can attract and concentrate radioactive particles formed from radon, a naturally occurring radioactive gas and known carcinogen.

They argue that this effect may cause more radioactive particles to be inhaled, sticking to the mouth, throat and lungs, from where they are absorbed, delivering an increased radiation dose to sensitive tissues such as the bone marrow and the fetus.

(from *Independent* 14 February 1996 with kind permission)

Identifying a cause and effect association between exposure to electromagnetic fields (EMFs) and cancer is not easy and, in spite of the news above, the link is yet to be confirmed.

Research into a possible association began in 1979, when a study suggested that children in the USA had an increased chance (two or three times more) of dying from cancer if they lived within 40 metres of overhead power lines. Of 14 further studies, eight have found positive links between EMF exposure and cancer, but results from other groups have been inconclusive.

In the USA, about 14 children in every 100 000 under 14 years old die each year from cancer. Of these 30% will die of acute lymphatic leukaemia, the commonest form of childhood cancer; there is only a 60% survival rate for patients with this cancer. But adults may also develop cancers; certain workers, such as those maintaining transmission lines, are exposed to quite high levels of EMF. Domestic electrical appliances are also known to emit EMFs but the health implications are not yet fully understood. Any increase in mutation rate in chromosomes caused by EMF exposure may lead to leukaemia and other cancers. So studies into the effects of EMF continue.

★ THINGS TO DO

1 Your body is exposed to ionising radiation from a variety of sources. Some of these are natural 'background radiation', as the table shows. Any further exposure (e.g. X-rays) increases the risk of mutation, and the bigger the dose the greater the risk.
   a) Why is the annual collection of this data important?
   b) Which sources expose people to higher than average doses?
   c) Why shouldn't pregnant women have X-ray treatment?
   d) Is the threat to human health from nuclear power stations greater than the benefits of the electricity generated? Find out about the arguments for and against the use of radioactive sources, then write an article to express the facts as you see them. Discuss your ideas with others in a group.

Source	% activity of radiation
radon gas	47
thoron	4
medical sources	12
rocks and buildings	14
food and drink	12
cosmic radiation	10
atmospheric fallout	0.4
job-related sources	0.2
nuclear waste	0.1
other sources	0.3

Source: NRPB

2 If possible, grow seeds that have been treated with mutagenic agents (such as X-rays) and compare them with untreated seeds. Try to investigate how they respond to different growth conditions.

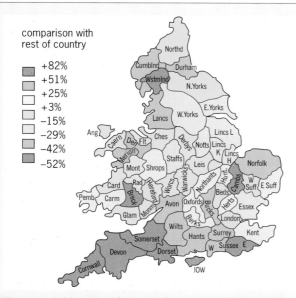

Malignant melanoma of skin in females (1968–1985) (from Swerdlow and Santos Silva *Atlas of Cancer Incidence in England and Wales, 1968–1985*, with kind permission of Oxford University Press)

3 Prolonged exposure to strong sunlight is now known to increase the rate of mutation in skin cells and lead to skin cancer (melanoma). The graph shows recent changes in the incidence of melanoma in Britain.
  a) Describe the distribution of the disease.
  b) What factor(s) could be associated with its incidence?
  c) How could a cause-and-effect relationship for melanoma be investigated?
  d) How could you investigate the 'sun-blocking' effect of different suntan creams?
  e) Find out more about how mutagenic agents can lead to cancers and other diseases.

4 New causes of cancer? A very-low-frequency electromagnetic field (EMF) is set up when electricity flows through electric power lines and appliances in the home. Such EMFs are a form of ionising radiation. Scientists are now investigating a possible link between this type of EMF and cancers such as leukaemia.
  a) How might this type of radiation cause leukaemia?

b) The data show the EMFs, measured in mG (milligauss), around three types of hairdryer. (b = background level of EMF from all sources.)

Hairdryer	Size of EMF (mG) distance from hairdryer			
	15cm	30cm	60cm	120cm
A	1	b	b	b
B	250	1	b	b
C	650	50	10	1

Identify three different variables that could expose a person to any harmful effects of EMF set up when a hairdryer is used.
  c) A cluster of leukaemia victims living in an area with many electric power lines may suggest an association. What would scientists need to do to prove that an EMF was causing leukaemia?

Exam questions

1 The chart gives some information about animals with backbones.

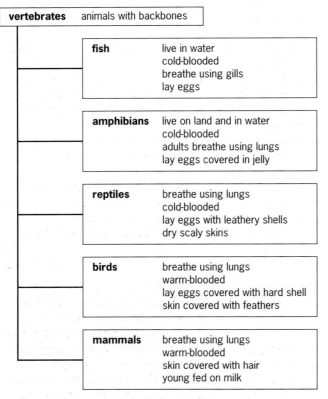

Use the information to anwer these questions.
a) Write down the name of the vertebrate group that breathes using gills. (1)
b) Write down the name of the vertebrate group in which the animals breathe using lungs and feed their young on milk. (1)
c) Describe **one** difference between amphibians and reptiles. (1)
d) Write down **two** features common to fish, amphibians and reptiles. (2)
(MEG, 1995)

2 All pupils in a class measured their height. The results are shown in the histogram.
a) Complete the table by showing the number of pupils for each height in the class.

Height (cm)	Number of pupils
130–139	1
140–149	
150–159	10
160–169	
170–179	
180–189	

(4)

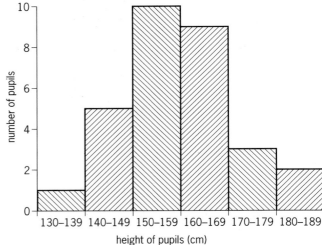

b) Why are the pupils in the same class different heights? Suggest **two** quite different reasons. (2)
c) Complete the sentences below.
The commonest range of heights is .... cm to .... cm.
This is known as the .... (3)
(NEAB, 1995)

3 The graph shows changes in the levels of three hormones in a menstrual cycle.
a) What does the graph suggest about how hormones might affect the release of eggs? Answer as fully as you can. (3)
b) One type of contraceptive pill keeps the level of hormone 2 high for most of the cycle. Suggest how this might work. (2)
c) Outline **two** arguments for and **two** against using hormones as contraceptives. (4)

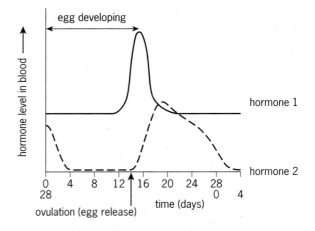

(NEAB, 1998 (specimen))

194

REPRODUCTION, INHERITANCE AND EVOLUTION

4 a) The drawings below represent seven stages of mitosis.

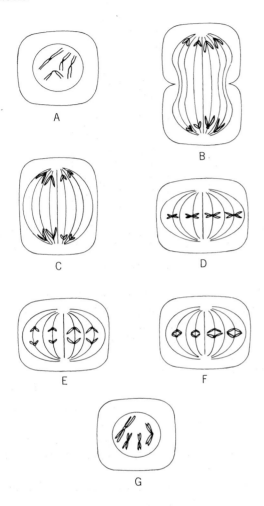

The drawings are **not** in the correct order of what happens to the chromosomes during mitosis. Rearrange them so that they are in the correct order.
The parent cell with four chromosomes is shown in drawing **A**. (2)

correct order **A** ....

(SEG, 1994 (part))

5 Busy Lizzies are small plants which are easily grown and produce large numbers of white, pink or red flowers all through the summer.
a) In Busy Lizzies the allele for red flowers (**R**) is dominant to the allele for white flowers (**r**).
A red-flowered Busy Lizzie (**RR**) was *cross-pollinated* with a white-flowered Busy Lizzie (**rr**). The seeds produced by this cross were grown to produce an F1 generation of plants all with red flowers.

i) What is meant by *cross-pollination*? (2)
ii) Why did **all** the F1 generation of plants have red flowers? (2)
iii) Two of these F1 red-flowered plants were cross-pollinated. The resulting seeds were grown to produce an F2 generation of plants.
Use a genetic diagram to explain the genotypes and phenotypes of the F2 plants.
You will be awarded up to **two** marks for the clarity of your genetic diagram. (6)
b) Imagine that on a remote, uninhabited island scientists discovered colonies of Busy Lizzies growing and that most of these plants produced **blue** flowers.
These plants could have evolved as a result of *natural selection* in action on a *mutant* blue-flowering Busy Lizzie plant.
i) Explain how a *mutant* (genetic mutation) can arise by referring to the way DNA replicates. (3)
ii) Explain how *natural selection* could have produced these colonies of **blue**-flowering Busy Lizzie plants. (3)

(SEG, 1995)

6 a) The Peppered Moth has light-coloured wings "peppered" with dark spots and bars. In 1850 a dark variety of this moth, with almost black wings, was discovered in Manchester. After 1850, the dark form rapidly became more and more common so that today, in the Manchester area, the light-coloured variety is rare. The light-coloured variety still continues to thrive in some other parts of the United Kingdom.
i) Suggest how the original dark-coloured variety came into existence. (2)
ii) **Explain** why the dark-coloured variety has now become the common variety in the Manchester area. (2)

(SEG, 1994 (part))

Glossary

Absorption The movement of food, gases or other chemicals into cells or into body fluid.

Active transport The movement of chemicals into or out of cells, needing energy from respiration.

Adenosine triphosphate (ATP) A chemical made during respiration as a temporary energy store.

Adrenal gland A gland above each kidney that makes the hormone adrenalin.

Adrenalin The hormone that increases heartbeat rate and breathing to prepare the body for vigorous exercise.

Aerobic respiration The release of energy from food by reacting with oxygen, producing ATP.

Aldosterone A hormone that allows the kidneys to balance salt loss with the body's need for salt.

Alleles Genes at the same site on a chromosome pair with information about one characteristic.

Alveolus Air sacs in the lungs where gases are exchanged between air and blood.

Amino acid The general name for about 20 types of chemical that combine to form different proteins.

Amniotic fluid Liquid that surrounds a developing fetus inside the uterus.

Anaerobic respiration The release of energy from food without the use of oxygen.

Antibody A chemical made by white blood cells, whose molecules have a shape that can recognise and inactivate foreign bodies.

Anti-diuresis A condition where concentrated urine is made by kidneys as more water is reabsorbed into blood from kidney tubules.

Aqueous humour A clear watery liquid in the eye.

Artery A type of blood vessel with a thick elastic wall, carrying blood *away* from the heart.

Asexual reproduction The production of identical offspring from one parent.

Assimilation The production of large complex molecules from smaller ones.

Atrium The smaller chamber of the heart.

Auxin A growth hormone in plants.

Bile The liquid made by the liver to reduce the size of fat droplets in the small intestine.

Biodiversity The range of different types of living things in an ecosystem.

Biological control The use of living things to prevent an increase in the population of another living thing.

Biological oxygen demand (BOD) A measure of the amount of oxygen used up by microbes as they digest organic matter in a sample of water.

Blind spot A disk of the retina of the eye that lacks light-sensitive cells.

Bowman's capsule The cup-shaped start of a kidney tubule, which filters liquid out of blood.

Bronchiole A smaller tube in the lungs through which air passes to and from alveoli.

Bronchus A wide tube connecting the trachea (windpipe) to a lung.

Callus A clump of unspecialised cells that grows when a plant is wounded and used for cell culture.

Cambium Plant tissue that divides to form new phloem and xylem.

Canine tooth A spear-shaped tooth, enlarged in carnivores, which can pierce and hold prey.

Capillary The smallest type of blood vessel that joins the smallest arteries to the smallest veins.

Carbohydrate Group of food chemicals, including starch, glycogen, cellulose and sugars, used as an energy source.

Carnivore An animal that eats other animals.

Cellulase An enzyme that digests cellulose.

Cellulose A complex carbohydrate used to make plant cell walls.

Cerebellum The part of the brain that coordinates muscle action.

Cerebrum The part of the brain that coordinates conscious action, including thought and memory.

Chlorofluorocarbons (CFCs) Chemicals used in some fridges and aerosols that destroy ozone.

Chlorophyll A green pigment in plant cells and some bacteria that absorbs light.

Chloroplast A part of a plant cell (an organelle), packed with chlorophyll.

Choroid layer A black layer in the back of the eye.

Chromatid One of 2 identical copies of a chromosome in a cell that is ready to divide.

Chromosome Part of a cell nucleus, which carry genetic information.

Ciliary muscle Muscle attached to the lens; it contracts and relaxes to change the lens shape.

Clone A group of identical cells or organisms.

Codon A sequence of three adjacent nucleotides on a DNA molecule that specifies an amino acid.

Cone Cell in the retina sesitive to colour.

Cornea A clear, curved outer part of the front of the eye.

Corpus luteum The remains of a follicle in a human ovary where progesterone is made.

Deciduous A type of tree that loses its leaves in autumn to reduce water loss.

Dentine A hard layer of a tooth, beneath the enamel.

GLOSSARY

Deoxyribonucleic acid (DNA) A complex molecule making up genes in cells.
Diffusion A passive movement of chemicals from an area of high to one of low concentration.
Digestion The chemical breakdown of large food molecules into smaller ones.
Diploid number The total number of chromosomes in a cell where there is a pair of each type of chromosome.
Diuresis A condition where the kidneys make a lot of dilute urine, because less water is reabsorbed.
Dominant trait An inherited feature that always shows in an offspring.
Dormancy A period of lower activity of an organism.

Ecosystem An interacting network of living things and their physical environment.
Effector A muscle or hormone gland that responds to nerve impulses following a stimulus.
Egestion Removal of waste faeces from the anus.
Embryo Ball of cells derived from a zygote.
Enamel The hard outer layer of a tooth.
Endosperm A food store in certain seeds.
Enzyme A chemical, made of protein that speeds up (catalyses) a reaction, but is unchanged itself.
Etiolation Rapid growth of plant stems, and a yellowing of leaves from lack of light.
Eutrophication Rapid growth of algae and other green plants in a water habitat.
Evolution The change in the type and number of living organisms on planet Earth.
Excretion The removal of waste materials made by cells in an organism.
Extracellular fluid A liquid derived from blood plasma that surrounds cells.

Fatty acid A chemical made of carbon, hydrogen and oxygen atoms, joined into long chains.
Feedback control A mechanism that switches off the release of hormones once their message has been acted upon.
Fertilisation A female sex cell (gamete) fusing with a male sex cell (gamete) to form a zygote.
Fetus The name for the human embryo once it has developed a human form.
Filtration The separation of fluid from blood into the start of a kidney tubule (nephron).
Flaccid The state of a cell that has lost water.
Follicle A sphere of cells that develops inside a human ovary and forms an egg inside.
Follicle-stimulating hormone (FSH) A hormone made by the pituitary gland, which stimulates growth of follicles in ovaries.

Gall bladder A sac attached to the liver, which stores bile.

Gamete A female or a male sex cell.
Gene A part of a chromosome, made of DNA and protein, which carries information to make one type of protein.
Genetic engineering A range of techniques used to change genes artificially.
Genome The total genetic make-up of a cell or organism.
Glucagon A hormone made by the pancreas that tells the liver to release sugar into blood.
Glucose A simple carbohydrate, a type of sugar, used as a source of energy in respiration.
Glycerol A chemical that is combined with fatty acids to make fat.
Greenhouse effect The steady accumulation of gases in the atmosphere, leading to an increase in world temperature.
Gross primary productivity The total amount of energy stored in plants in an ecosystem, as the result of photosynthesis. Measured as $kJ\ m^{-2}\ y^{-1}$.
Guard cell One of two curved cells on the surface of a leaf surrounding a stoma.

Habitat A place where plants and animals live.
Haemoglobin A chemical in red blood cells that carries oxygen; consists of globin plus iron.
Haploid number The total number of chromosomes in the cells of an organism where only one set of chromosomes is present.
Herbivore An animal that eats only plants.
Heterozygous A genome with two different alleles (one dominant, one recessive) for a characteristic.
Hibernation An animal in a resting state to conserve energy over winter.
Homeostasis Processes in the human body that keep internal conditions steady.
Homologous chromosomes Pairs of chromosomes that have genes for the same characteristics.
Homozygous A genome with two identical alleles for a characteristic.
Hybrid Offspring formed when gametes combine from different species, races or varieties.
Hypothermia Excessive cooling of the body of a bird or mammal that can cause death.
Hypothesis A plausible guess, based on everyday or scientific knowledge, that can be tested by investigation.

Implantation The attachment of an embryo on to the uterus wall.
Incisor A flat, cutting tooth at the front of the jaw.
Inoculation Giving vaccines to a human or animal to improve resistance to disease.
Insulin A hormone made by the pancreas that tells the liver to absorb sugar from blood.

GLOSSARY

In-vitro fertilisation Artificial fertilisation of an egg by a sperm outside the body.
Iris The coloured disk in the front of the eye, which changes shape to widen or narrow the pupil.
Large intestine The wide, second part of the intestine where water is absorbed into blood.
Lens Part of the eye that focuses light on to the retina.
Leucocyte A type of white blood cell that makes antibodies or antitoxins.
Ligament A slightly elastic tissue that joins bones.
Lignin A substance that strengthens xylem vessel walls and makes wood.
Lincoln index A formula used to calculate an animal population size.
Luteinising hormone (LH) A hormone made by the pituitary gland that increases oestrogen.
Lymph A fluid formed from blood plasma, which returns substances to the blood.
Marrow The tissue inside larger bones where white blood cells and platelets are made.
Medulla oblongata A region of the brain controlling heartbeat rate and breathing.
Meiosis A 'reduction' division; type of cell division where a diploid cell produces up to four haploid cells, prior to gamete formation.
Membrane The outer part of a cell that controls the passage of most substances in or out.
Menstrual cycle The monthly sequence of changes in a female's body, which includes egg release, thickening of the uterus wall and periods.
Microorganism Microscopic living thing, also called a microbe, includes bacteria and viruses.
Mitochondrion A structure (organelle) inside a cell where most reactions of aerobic respiration occur, producing ATP.
Mitosis The commonest type of cell division, where a 'parent' cell divides into two 'daughter' cells, with identical chromosomes.
Molar A grinding tooth in the back of the jaw.
Mutation A sudden change in a gene or chromosome, usually having a harmful effect.
Myelin sheath A layer of fatty material around a nerve cell fibre that speeds up impulses.
Nephritis A disease causing kidney inflammation.
Nephron A microscopic tubule in a kidney.
Neurone Also called nerve cell; cell specialised to carry impulses from and to adjacent cells.
Node A region of a plant stem where a leaf stalk attaches.
Nucleotide A compound consisting of a sugar plus a base (nucleoside).

Oestrogen A hormone made by the ovaries that produces secondary female sex characteristics, prevents release of FSH but stimulates LH release.
Omnivore An animal that eats plants and animals.
Optic nerve The nerve that carries impulses from an eye to the brain.
Osmosis Passive movement of water across a selectively permeable membrane.
Osteoporosis Disease caused by mineral salt absorption into blood from bones.
Ovary An organ in a female mammal where eggs and sex hormones are made.
Oviduct The tube connecting an ovary to the uterus.
Ovulation The release of eggs from ovaries.
Ovule Part of an ovary in a plant, which contains the female gamete.
Ozone A form of oxygen (O_3) formed from O_2.
Pancreas An organ that makes pancreatic juice and the hormones glucagon and insulin.
Pepsin Enzyme that catalyses digestion.
Peristalsis Repeated contraction and relaxation of muscles in the gut wall that moves food.
Phloem A tissue in a plant through which foods and hormones travel to all parts of the plant.
Photosynthesis Process in a green plant or some bacteria, using light energy, carbon dioxide and water to make carbohydrate and oxygen.
Pituitary body Organ attached to the brain that makes a number of hormones.
Placenta An organ that enables materials to be exchanged between the mother's blood and embryo.
Plasma The watery liquid part of blood.
Platelet Blood cell type responsible for clotting.
Plumule The part of a seed that grows into a shoot.
Pollen A plant structure that carries the male gamete.
Premolar A crushing tooth with a flat crown.
Progesterone A female sex hormone that stimulates the uterus wall to thicken.
Protozoan A single-celled, microscopic animal, such as *Amoeba*.
Pulp cavity A space in the centre of a tooth filled with living cells, nerves and blood vessels.
Pupil The hole in the middle of the iris.
Pyramid of mass/number A diagram representing the total mass/number of organisms at each stage of a food chain.
Quadrat A square frame used to sample organisms in an area.
Radicle The part of a seed that grows into a root.

GLOSSARY

Recessive trait An inherited feature that shows in an offspring only when both alleles are present.

Recombinant chromosome A chromosome that has been changed by a part of it being swapped for a part of another chromosome.

Rectum The end of the digestive system where waste faeces is stored prior to egestion.

Red blood cell A biconcave-shaped blood cell that carries oxygen.

Reflex arc The simplest connection of nerve cells, giving an automatic, rapid response

Replication The doubling up of DNA to make two identical copies of a chromosome (chromatids).

Respiration The production of ATP, which stores energy released from food.

Retina The layer of light-sensitive cells in the eye.

Rod A cell in the retina of the eye, sensitive only to white light from low to high levels of intensity.

Runner A stem-like part of a plant from which grow newly formed plantlets.

Saliva A liquid that lubricates food in the mouth.

Salivary glands Organs that make saliva and release it into the mouth.

Sclera The tough, white outer layer of the eye.

Secondary thickening Sideways growth of roots and stem, increasing thickness.

Selective breeding The production of offspring with desired characteristics by breeding (crossing) parents with suitable features.

Sense organ An organ containing sensory cells.

Sensor Part of an organism that can detect changes (stimuli).

Sexual reproduction The production of offspring from the joining of gametes from sex organs.

Small intestine The narrow first part of the intestine where food is digested and absorbed.

Sperm A male gamete, made in a testis.

Spinal cord Part of the nervous system; connects most nerves from the body to the brain.

Stamen The male part of a flower.

Stigma Part of the flower on which pollen lands.

Stoma (stomata = plural) A hole in a leaf surface through which gases can enter and escape.

Style Part of a flower connecting the stigma to the ovary.

Suspensory ligament A ring of tissue around the lens in an eye, pulled by ciliary muscle contraction.

Synapse The tiny gap between neurones.

Synovial fluid A liquid that lubricates bones sliding against each other in a joint.

Synthesis Chemical reactions that make large, complex molecules from smaller, simpler ones.

Taste bud A group of cells on the tongue sensitive to chemicals.

Tendon Tissue joining a muscle to bone.

Testis Male reproductive organ that makes sperm.

Tissue A group of similar cells, e.g. muscle tissue.

Tissue (cell) culture A range of techniques used to grow cells in artificial conditions outside the body of a plant or animal.

Trachea The windpipe; a tube strengthened by cartilage rings between the lungs and nose/mouth.

Translocation (two meanings): movement of food through phloem tissue in a plant; change in chromosome structure caused by exchange between chromosomes.

Transpiration Loss of water from leaves and stem of a plant.

Transpiration stream The flow of water in a plant, through xylem tissue, from roots to leaves.

Tropism Plant growth affected by certain stimuli.

Tuber A root or stem of a plant that stores food.

Turgid The state of a cell when swollen with water.

Urea Poisonous waste chemical, excreted in urine.

Uterus Part of the female reproductive system where an embryo develops (womb).

Vacuole A space inside a cell filled with solution.

Vagina A part of the female reproductive system.

Vascular bundle A group of transport tissues in a plant.

Vasodilation The widening of blood capillaries.

Vegetative propagation A type of asexual reproduction in plants.

Vein A blood vessel carrying blood *back* to the heart.

Ventricle The largest chamber of the heart.

Villus A tiny projection of the small intestine wall.

Virus A simple type of microbe, consisting of DNA or RNA surrounded by a protein coat.

Vitreous humour A clear, colourless jelly in the eye.

White blood cell Type of blood cell that has a nucleus and fights disease.

Womb see uterus.

Wood The tough, fibrous cellular substance in trees and shrubs, consisting largely of cellulose and lignin.

Xylem The tissue in a plant through which water and minerals pass from roots to other parts.

Zygote A fertilised egg formed as a male and female gamete combine.

Index

Page numbers in **bold** type show where a subject is most fully explained.

acid rain 141
active transport 93
   in kidney 36
adaptation 106–7
addiction 55–7
ADH (anti-diuretic hormone) 36
adrenal glands 43, 45
adrenalin 45
aerobic respiration 18
   in plants 82
air pollution 140–1
alcohol 54, **56**, 57, 60
   risks to fetus 177
alleles 164
   dominant/recessive 184–5
   inheritance of diseases 186–7
   selective breeding 174, 175
alveoli 20, 21
amino acids 14
   formation 166
amylase 15, 17
anaemia 29
   sickle cell 187
anaerobic respiration 19
   in plants 83
Animal Kingdom 152
animal welfare 117, 122–3, 145
anorexia nervosa 10
antibiotics 190
   resistance to 191
antibodies 32, 33, 60
antigens 60
arteries 27
artificial worlds 146
asexual reproduction 162, 163
   in plants 162, 168–9
asthma 140, 141
astronauts 5
athletes *see* sportspeople
ATP (adenosine triphosphate) 16, 18, 82
auxin 94, 95
   in rooting powder 96

bacteria 30
   against pollution 137
   antibiotic-resistant 191
   earliest 156
   genetic engineering and 188
   nitrifying 131
   reproduction 162
balance 48, 49
banana storage 96
battery farming 117
benzene pollution 140, 141
bile 17
biodegradable 124–7
biological control 119, 134–5
Biosphere-2 146
birds 152
   dodo 154
   finches 158
   sparrowhawks 113, 133
blind spot 51
blood **28–9**
   circulatory system 3, **26–7**
   clot 29, **31**
   composition 29
   groups 29
   transfusion 28–9
   vessels 27, 38
bones 3, **6–7**
   broken 5, **6**
   fossil 155, 156
   osteoporosis 42
brain 47, **52**
breathing 3, **20–1**
   at high altitude 20
   gas exchange 19

caffeine 54
calcium 42, 43
   in cell walls 80
   in diet 10–11
camouflage 158–9
cancer 191
   causes 192, 193
   leukaemia 191, 192, 193
   skin cancer 140–1, 193
capillaries 27
carbohydrates
   digestion of 14, 16
   in diet 10
   in plants 70–1, 74
carbon cycle 128–9
carbon dioxide
   from respiration 18–21
   in air 128
   in global warming 129
   in photosynthesis 70–1, 74, 76
carnivores 112
   carnivorous plants 79
   teeth of 9
cartilage 6
cells 2–3
   blood 29, 31, **32**
   bone 7
   cell culture 171
   division (meiosis) 162–3, **165**
   division (mitosis) 162–3, **164**
   muscle 4
   nerve 53
   plant 62, 63
cellulase 12
CFCs 141
chicken farming 117
chlorophyll 70
chloroplasts 62, 70
chromosomes 160, 162, **164–5**
   alleles 164, 184–5
   gene mapping 167
   sex chromosomes 164, 182
   translocated 191
cigarettes 54, **56**, 57
   risk to fetus 177
circulatory system 3, **26–7**
   blood **28–9**
   temperature control 38–9
classification systems 108–9, **152–3**
climate 102, 103
   global warming 103, **128–9**, 147
clones/cloning 162, **170–1**
clotting of blood 31
cold-blooded animals 41
colour blindness 183
colour vision 52
communities 118–19
competition between species 118
compost 126, 130
cones (eye) 52
coniferous trees 68, 106
conservation 142–3
consumers 112
contraception 180–1
cystic fibrosis 186

Darwin, Charles 158, 159
DDT 133
decay 124–7
   compost makers 126
deciduous trees 68, 106
dentine 8
diabetes 35, **44**, 45
   insulin production 188
dialysis 37
diet **12–13**
   diet tables 10–11
   heart disease and 22
diffusion 21, 92
   in lungs 21
   in plants 92
   in small intestine 14
digestion **14–17**
   digestive enzymes 14, 15, **16–17**
   digestive system 3, 15
   saliva 8
dinosaurs 154, 156
diseases **30–3**
   causes 30–1
   cystic fibrosis 186
   fighting disease 32–3
   heart disease 23
   Huntington's chorea 187
   inheritance of 186–7
   kidney disease 36–7
   leukaemia 191, 192, 193
   liver cirrhosis 56, 57
   lung cancer 56, 57
   sickle cell anaemia 187
   skin cancer 140–1, 193
distribution studies 110–11
DNA 160, 164, **166**
   fingerprinting 189
   mutations 190–3
dodo 154
double helix 166
Down's syndrome 164, 167
drugs 54–7
   alcohol 54, **56**, 57, 60
   drug addiction 55–7
   fertility drugs 179
   penicillin 190
   risks to fetus 177
dung
   as energy source 126
   organisms on 125

ear 48, **49**
   number of bones 6
Earth (formation of) 154, 156
earthworms 7, 102
   counting 111
eating disorders 10
ecosystems 102, 142–5
   artificial 146
   carbon cycle 128–9
   computer modelling 147
   decay and 124–5
   fish aquarium 146
   nitrogen cycle 130–1
   working with 142–5
egestion 34
enamel 8
endangered species 103
endosperm 65
energy
   requirements 10, 11
   sources 126, 144
   transfer 116–17
enzymes **16–17**
   cellulase 12
   digestive 14, 15, **16–17**
etiolation 69
eutrophication 130–1
evolution 154, **158–9**
excretion **34–7**
   urinary system 3, 35–6
exercise 4, **5**, **22–3**
   anaerobic respiration 19
   athletes at risk 10
   blood sugar levels 15
   heart rate and 25
   heart size and 23
exhaust fumes 140, 141
extinction of species 103, 154, 157
   fossils **154–7**
eye **50–1**, 52

INDEX

family planning 180–1
farming methods 145
   battery farming 117
   fish farming 150
   selective breeding 175
fats
   digestion of 16
   fatty acids 14, 16
   in diet 11
feedback control 45, 181
fermentation 83
ferns 153
fertilisation 162, 176
   in-vitro 179
fertilisers 78, 80–1, 99, 130
fertility drugs 179
fertility rates 121
fetus 177
fever 40
fibre in diet 11, **12**
fish
   aquarium 146
   farming 150
   stocks 143
fitness 22
flowering plants 153, 172–3
fluoride 9
food
   banana storage 96
   decaying 125
   modern food plants 174
   nitrate levels 131
   see also diet
food chains **112–13**, 114, 148
   DDT accumulation 133
   trophic levels 116
food webs 114–15
fossils **154–7**
   fossil fuels 141, 144
free range farming 117
FSH (follicle-stimulating hormone) 179, 180, 181
fuels 141, 144
fungi
   fungal diseases 30, 85
   on decaying food 125
   useful 190
   yeast 83, 163

Gaia hypothesis 147
gall bladder 17
gametes 162
genetics **160–1**
   boy or girl? 182
   gene mapping 167
   genetic code 160–1
   genetic counselling 186–7
   genetic engineering 188–9
   inheriting features 183, **184–5**
   mutations 190–3
   plant variation 172
   selective breeding 174–5
   twins 182
   variation 160–5
geotropism 94

germination 65, **66**, 67
germs 30
global warming 103, **128–9**, 147
glucagon 44–5
greenhouse effect 103, 128, **129**
growth 2
   plant 68–9, 94–7
gut (digestive system) 3, 15

habitats 102, 106
   destruction by humans 102, 122–3, 145
   food energy and 116
   in Britain 104
   saltmarsh 107
   studying 110–11
   survey on 105
   woodland 106
haemoglobin 29
haemophilia 31
hearing 48, **49**
heart **24–5**
   beats per year 22
   heart disease 23
   heart muscle 4, 25
   keeping healthy 22
height/weight charts 161
hepatitis 35
herbivores 12, 112
   teeth of 9
heterozygous 184, 185
hibernation 41
homeostasis 42–3
homozygous 184, 185
hormones **43**
   ADH (anti-diuretic) 36
   adrenalin 45
   contraceptive 180–1
   female sex hormones 43, 179, 180–1
   fertility 179
   FSH (follicle-stimulating) 179, 180, 181
   in blood 29, 43
   in plants 94, 95
human evolution 155
human genome project 167
human population 121
human reproduction 176–81
   boy or girl? 182
   contraception 180–1
   planned pregnancy 178–9
   twins 182
humus 129
Huntington's chorea 187
hybrids 174
hydrotropism 94
hypothermia 40

identification systems 108–9, **152–3**
immune system **32–3**, 60
   transplantation and 36, 188
   white blood cells 29, 31, **32**

immunisation 33
inheritance **184–5**
   mutations 190–3
   of colour blindness 183
   of diseases 186–7
   of eye colour 194
   of tongue rolling 184–5
   see also genetics
insecticides 132–3
insulin 44–5
   by genetic engineering 188
invertebrates 152, 156
iron in diet 10, 11

Jenner, Edward 33
joints 6

keys (biological) 108–9, 152–3
kidneys 34, **35–7**
Kingdoms 152–3

lactic acid 19
land reclamation 137
Latin names 153
lead pollution 136, 137
leaves
   arrangement 72
   decay of 124
   features 74, 99
   flow of chemicals 99
   photosynthesis 70–7
   starch test 73
   water loss 90–1
leukaemia 191, 192, 193
life-cycles 124–5
ligaments 6
light meter 110
light response 94–5
lignin 86
Lincoln index 111
Linnaeus, Carolus 153
liver 44–5
   cirrhosis 56, 57
lung cancer 56
lymph 14

mammals 152, 156, **160**
manure 134
   dung as energy source 126
   green manures 135
   organisms on cow pat 125
marrow 6
medicines 54, 190
meiosis 162, **165**
Mendel, Gregor 175
menstrual cycle 176, **180–1**
microorganisms 30–2
   decay due to 124–5
   useful 131, 137, 190
microscopes
   leaf sections 75
   leaf surfaces 91
   plant cells 63
   pollen grains 173
   yeast budding 163
milk 13

minerals
   calcium 42, 43
   fluoride 9
   for bone development 7, 42
   for plant growth 71, **78–81**, 93
   in diet 10, 11
mitochondria 4, 82, 93
mitosis 162–3, **164**
movement
   muscle system 3, **4–5**
   plant phototropism 94–5
muscles 3, **4–5**
muscular dystrophy 191
mutagens 190, 191
mutation 190–3

natural selection 158–9
   mutations and 191
nephritis 36
nephrons 36
nerve cells (neurones) 53
nervous system 3, **46**
   brain 47, **52**
   cells 53
   reaction times 46–7
nicotine 54, 56
   risks to fetus 177
nitrogen 60
   nitrogen cycle 130–1
   nitrogen oxides 140, 141
nuclear fuels 144
nutrients for plant growth 71, **78–81**, 102
   recycling 124, 128–31
nutrition 2, **10–17**
   diet requirements 10–13
   see also digestion

oestrogen 43, 180, 181
oil spillages 138–9
omnivores 12, 112
organic growing 134–5
organ systems 3
osmosis 89, **92**
osteoporosis 42
otters 133
ovaries 43
owls 106
ozone 140, 141

pancreas 44–5
peat bogs 129
penicillin 190
pepsin 17
peristalsis 12
pest control
   biological 119, 134–5
   pesticides **132–3**, 134
pH
   enzyme action and 17
   in mouth 8
   of blood 43
phloem 63, 86, 87
phosphate pollution 136
photosynthesis **70–7**, 116

INDEX

phototropism 94–5
physical exercise *see* exercise
pituitary gland 43, 52, 180
placenta 177
plants
  as energy source 144
  cloning 170, 171
  energy efficiency 116
  grassland 109
  growth 68–9, 94–7
  herbicides 132
  life cycles 64
  mineral needs 71, **78–81**
  organisation 63
  photosynthesis **70–7**, 116
  plant cell 62, 63
  Plant Kingdom 152, 153
  pollination 118, 172–3
  propagation 168–9
  reproduction (asexual) 162, 168–9
  reproduction (sexual) 172–3
  respiration 82–3
  seed banks 157
  selective breeding 174
  soil preferences 104
  vegetative propagation 162
  water needs 84–5
  woodland 106
plaque 8
plasma 29
platelets 29, 31
pollination 172–3
  by insects 118, 173
pollution **136–9**
  air 140–1
  nitrate 130
  water quality control 138
population studies **120–1**, 123
potatoes 65, 67, 169
potometer 90
predators and prey **118–19**, 120, 135–6
pregnancy 176–7
  planned 178–9
  preventing 180–1
producers 112
proteases 16
  pepsin 17
proteins
  digestion of 14, 16
  formation 166
  in diet 10, 11
  nitrogen cycle 130–1
protozoa 30
  Toxoplasma gondii 31
pyramid of mass/numbers 115

quadrats 110–11

radiation hazards 192
radioactivity hazards 190, 191
reaction times 46–7

recycling 127, 137
  carbon cycle 128–9
red blood cells 29
  in urine 35
  sickle cell anaemia 187
reflexes 46–7, 53
reproduction **162–3**
  asexual 162, 163
  sexual 162–7, 172–3, 176–7
  see also human reproduction
respiration **18–19**
  breathing system 3, **20–1**
  in plants 82–3
retina 50, 51
rods (eye) 52
roots 62, 79, **88–9**, 97
  geotropism 94
  hydrotropism 94
  root nodules 131
  rooting powder 96, 168
  water entry 89, **92**
  waterlogged 82
rubbish recycling 127
rubbish tips 126

saliva 8
  investigations 15, 17
saltmarsh 107
salts
  in blood 29
  salt balance 35
seal populations 120
seas
  overfishing 143
  seawater pollution 138
seeds 65–7
  seed banks 157
selection *see* natural selection
sense organs 46–51
  eyes 50–1
sewage treatment 127
sex hormones 43, 179, 180–1
sexual reproduction 162–7
  human 176–81, 182
  plants 172–3
  selective breeding 174–5
sickle cell anaemia 187
skeleton 3, **6–7**
  broken bones 5, **6**
  fossil bones 155, 156
  osteoporosis 42
skin 38, **48**
  skin cancer 140–1, 193
  touch sense 49
sleeping sickness 30
slug traps 135
smallpox 32–3
smell sense 48
smoking 54, **56**, 57
  risks to fetus 177
soil 102, **104**
  carbon levels 129
  land reclamation 137

nitrogen cycle 130–1
  organic gardening 134–5
  pollution 136
  sampling 110
solvent abuse 55
sparrowhawks 113, 133
sperm 176–9, 182
sportspeople 22–3
  athletes at risk 10
  drug testing 54
  heart rate 25
  heart size 23
  marathon runners 39
squirrels 118
starch 15
  test 73
stems 86–7
  phototropic 94–5
steroids 54, 60
stomata 90–1
succulents 85
sugar control 43
  diabetes 35, **44**, 45
sugars in plants 75
sulphur dioxide 140, 141
surveying populations 110–11
sweating 38–9
synapses 53
synovial fluid 6

tadpoles 106, 171
taste buds 48
teeth **8–9**
  tooth decay 8
temperature
  control in humans 38–9, **40–1**
  of body 38–9
  photosynthesis rate 76
tendon 4
testes 43, 178
testosterone 43
thyroid gland 43
tissue culture 171
tissues
  animal 2–3
  plant 63
tongue rolling 184–5
toxins 31
toxoplasmosis 31
traffic exhaust fumes 140, 141
transfusion (blood) 28–9
translocation 191
transpiration 84
transplantation of organs 36, 188
transport
  active 93
  plant minerals 79, 93
  plant sugar 75
  plant water 84
trees 68, 106
  annual rings 69, 87
  bonsai 94
  fossil 157

nitrogen cycle 130–1
organic gardening 134–5
pollution 136
sampling 110
oak tree food web 114–15
transplanting 88
trypanosomes 30
tubers 65
twins 182

urea transport 29
urinary system 3, 35–6
urine 35
  production 26, **36**

vaccinations 33
variation 160–3, **164–5**
  plant 172
  selective breeding 174–5
vegetarianism 12
vegetative propagation 162
veins 27
vertebrates 152, 194
villi 14, 15
viruses 30–1
vision 50–1
  colour 52
vitamins 11
  vitamin A 10, 11
  vitamin B 11
  vitamin C 10, 11, 13
  vitamin D 7, 11

warm-blooded animals 40–1
water in humans
  balance 35
  loss through skin 38–9
water in plants
  balance 90–1
  germination and 66, 67
  hydrotropism 94
  in stems/stalks 86–7
  requirements 84–5
  transport 84, **88–9**
water quality control 138
weedkillers 97
weight/height charts 161
welfare of animals 117, 122–3, 145
whaling 142
white blood cells 29, 31, **32**
  leukaemia 191
whooping cough 33
wildlife protection 122–3
woods 106
wormery 126

X chromosome 182
X-ray hazards 190, 191, 192
xylem 63, 84, 86–7

Y chromosome 182
yeast 83
  reproduction 163

zygote 163